VALUATION:
Special Properties and Purposes

Edited by
Phil Askham

2003

Routledge
Taylor & Francis Group

LONDON AND NEW YORK

First published 2003 by Estates Gazette

Published 2014 by Routledge
2 Park Square, Milton Park, Abingdon, Oxon OX14 4RN
711 Third Avenue, New York, NY 10017, USA

Routledge is an imprint of the Taylor & Francis Group, an informa business

ISBN 978-0-7282-0418-8 (pbk)

© Note: For details as to copyright see end of each chapter

Typeset by Amy Boyle, Rochester, Kent

Preface

Many readers will be familiar with *Valuation: Principles into Practice,* edited by Bill Rees and Richard Hayward, one of the classic valuation texts, described as a handbook for students and practitioners. The fifth edition, published in 2000 runs to over 800 pages with its 20 plus chapters covering an astounding range of property types and valuation methods. Despite the size of this weighty tome a good deal of possible material was left out.

Bill Rees always felt that there was enough material left, as it were, on the cutting room floor, to warrant a follow up text and when he asked me if I would consider this I was delighted to accept the task. This volume, dealing with 10 additional topics, should be seen as a companion to *Principles into Practice.* As such it attempts to follow the same philosophy, providing a bridge between valuation theory and practice. It also, in Bill's words, represents a coming together of 'town and gown' in as much as the chapters are written by academics as well as practitioners, although the emphasis remains firmly on the practical.

It is appropriate at this point to make clear a number of caveats. Authors' opinions are confined to the chapter bearing their name and equally, as editor I take no responsibility for the opinions expressed. Each chapter is intended as a guide or starting point for further research and some of these represent very specialised areas of valuation which may well be beyond the experience of the average general practitioner. Legislation is taken to be that in effect at March 2003. Equally chapters will reflect economic and market conditions at that date.

Finally, I am greatly indebted to the authors whose professional responses to my brief left me with a relatively simple and a very pleasurable editing task.

Phil Askham

Contents and Authors

Table of Cases

Table of Statues

Table of Statutory Instruments

Chapter 1
Milk Quotas

Financial crisis is never lurking very far from the Common Agricultural Policy (CAP) of the European Union (EU). Its cause in the early 1980s was milk. Half the CAP budget was absorbed supporting milk and milk products. Several non-marketing and herd conversion schemes had failed to tackle the problem, price control was unacceptable politically and demand for milk and milk products was falling. Something had to be done. Production controls – quotas – were until the last minute rejected as a solution. But in late-March 1984 milk quotas were chosen as the solution to the European milk crisis. Milk quotas were introduced on 2 April 1984.

At first little was known about the detailed implications of milk quota, and dairy farmers found themselves facing an uncertain future. Would they have enough quota? How could more quota be acquired (*could* it be acquired)? Could quota be sold? Who 'owned' quota and how is it categorised legally? What about investment plans, and the position of new and retiring tenants? These questions dogged farmers and their advisers for several years after quota was introduced – indeed there are still aspects which lack a definitive answer today.

This chapter discusses the importance of milk quota to the practising valuer. It explains the history, legal status and statutory framework for milk quota, before considering the impact of quota on the sale and transfer of farmland. The impact of quota on valuations for other purposes is also considered. Finally some of the more specialised valuations which can be required in connection with milk quota are introduced with reference to further sources of information for the valuer who may wish to extend his or her expertise in this area.

Importance of milk quota to practising valuers

Any valuer who is called upon to value land which may have been used for dairy farming, or who is concerned with a valuation of the assets of a dairy farming business, should be aware that milk quota can be a very valuable asset. Even a modest dairy farm of 80 cows may be using milk quota worth £100,000 (500,000 litres of quota at 20p/litre). Entitlement to this asset may need to be apportioned between areas of land, and between landlord and tenant. But its availability can have a significant effect on the valuation of quota-bearing land for many different purposes, for example sale or taxation. Its availability can be a vital consideration in future planning: business restructuring, farm diversification, land rationalisation or compulsory purchase for example. Therefore, even the general practitioner who may only occasionally encounter rural valuation work should be aware of the implications of milk quota.

Legal and administrative framework

Milk quota works by setting a national limit for milk production. The national limit is shared between milk producers. This is the dairy farmer's milk quota. If national milk production exceeds the national limit, a penalty is payable by dairy farmers who have exceeded their quota. The penalty is called the Milk Supplementary Levy. It is used to fund the storage and disposal of surplus milk through various EU schemes for this purpose.

The framework for the regulation of milk quotas originates in European law. The current regulations are:

- Council Regulation (EEC) No. 3950/92 (as amended), which establishes an additional levy on milk and milk products.
- Commission Regulation (EEC) No. 1392/2001, which sets out detailed rules on the levy.

European regulations must be converted into national law in each member state, and the relevant regulations in the UK currently are in the following statutory instruments (SIs):

- SI 2002 No 457 The Dairy Produce Quota Regulations 2002 (England)
- SI 2002 No 88 The Dairy Produce Quota Regulations 2002 (Northern Ireland)
- SI 2002 No 110 The Dairy Produce Quota Regulations 2002 (Scotland)
- SI 2002 No 897 (W103) The Dairy Produce Quota Regulations 2002 (Wales).

In addition, valuers also need to be aware of the provisions in the Agriculture Act 1986 which govern the entitlement of some tenants under the Agricultural Holdings Act to compensation at the end of their tenancies. The Act is supplemented by SI 1992 No 1225 the Milk Quota (Calculation of Standard Quota) (Amendment) Order 1992.

Despite this regulatory framework, it has proved hard to define the nature of a milk quota in English law. Most commentators have settled on the right which it gives a dairy farmer to market milk without incurring a levy. It might be noted in passing that milk quota is fundamentally different from the other livestock quotas which came into being as a result of the 1992 reforms to the Common Agricultural Policy. The later quotas entitle farmers with sheep or suckler cow quotas to claim an annual subsidy payment from the Common Agricultural Policy (in practice by claiming from the UK's Rural Payments Agency).

Early administration of milk quotas

In early 1984, national quotas (also known as reference quantities) were allocated to member states and the member states allocated individual quotas to producers. In practice, not all of the national quota was allocated to individual producers because some was held back to form a national reserve. The initial allocation of quota to producers was based on their previous level of milk production. Quota is broadly divided into two types:

- Wholesale quota, is held by the majority of dairy farmers, and applies to milk which is sold to approved wholesale purchasers. The initial allocation of wholesale quota in April 1984 was based on 1983 production, less the 9% retention for the national reserve of quota.
- Direct Sales quota applies to milk and milk products which are marketed direct to the public. The initial allocation of this quota was based on 1982 production, less the allocation to the national reserve.

Quota, whether wholesale or direct, covers milk and milk products. Milk products include butter, cheese, yoghurt and ice cream. The quota covers all milk and milk products which leave the holding. Direct sales quota can be redesignated as wholesale quota, and wholesale quota can be redesignated as Direct Sales quota.

Farmers were able to appeal against their original allocation of quota, typically because their production had been lower than normal during the 'reference year' on which their allocation of quota was based, or because they were in the middle of an expansion programme. A successful appeal led to an award of additional quota from the national reserve. Another three categories of farmer were able to benefit from the later award of 'special quota', which is also known as SLOM quota. SLOM quota was awarded to farmers who were not in production in 1984 because they were participating in some of the earlier non-marketing or herd conversion schemes (SLOM is an acronym for the Dutch scheme). There were three allocations of SLOM quota in 1989, 1991 and 1993 to farmers who had participated in the various schemes that had attempted to curb milk production before 1984. The first and third allocations (SLOM 1 and SLOM 2) are now regarded as ordinary quota. Originally farmers were not allowed to transfer or lease out SLOM quota but those restrictions no longer apply. Nevertheless SLOM quota is still shown separately in official quota records.

Milk quota refers to annual milk production in the 'quota year' which runs from 1 April to 31 March. The original allocations of quota were expressed solely in litres. Farmers began to realise that they could earn greater profits within their quota by improving milk quality. This was achieved by raising butterfat levels (through feeding and breeding policies). This frustrated the intention of quotas to reduce the production of milk and milk products, as products like cream, butter and cheese are made from the butterfat in milk and surpluses in these products began to grow again. So Butterfat bases were added to wholesale milk quota in the 1987/88 quota year. For most producers the butterfat base was based on the average fat content of their milk in the 1985/86 year. Deliveries of wholesale milk are adjusted for quota purposes by reference to their butterfat base. Milk which is richer than the national average butterfat base is increased in volume for quota purposes, while milk which is weaker is reduced in volume. The national average butterfat base in May 2002 was 3.97%, and the current formula adjusts milk volume by 0.18% for every 0.01% variation in butterfat from the national base. This can mean that a farmer who has kept slightly within quota, but has exceeded his or her butterfat base, may nevertheless incur the milk supplementary levy because the adjusted milk volume exceeds his or her quota. Conversely a farmer who exceeds quota with weak milk may find himself within quota when the adjustment has been made.

It can be seen that an understanding of milk quota is crucial to the modern dairy farmer. It is understood that milk quotas will be in place until at least March 2008, although the other aspects of the CAP dairy regime are to be reviewed soon. Changes may include a levy on larger dairy farmers to pay for supporting the milk markets, a process known in CAP jargon as 'degressivity'. The ability to transfer quota, permanently or temporarily, is clearly important if any flexibility is to be retained in the dairy farming industry.

Transferring milk quota

The general principle of quota ownership is that is attached to a holding, not a producer. Generally therefore milk quota will go to the new occupier if a holding changes hands. However, quota and the holding on which it is held are not inseparable. Quota can be transferred permanently or temporarily, with or without a transfer of land.

Permanent transfer with land

Sale of land 'used for milk production'

The most straightforward way to transfer quota to another producer is to sell the land with which it is associated to that producer, with the agreement of all interested parties (landlords of other land with an interest in the quota, mortgagees and so on) and to notify the Rural Payments Agency.

Lease land 'used for milk production'

A lease is agreed under which the purchaser of the quota rents land from the vendor. The tenant/quota-buyer occupies the land, *but does not use it for milk production*. Instead he or she uses the quota to produce milk from another part of his or her holding (either by intensifying milk production on the other part or by using other land which has not previously been used for milk production). The lease must be at least 10 months long in England or Wales, only eight months in Scotland, but 12 months in Northern Ireland. This process is sometimes called 'massaging' quota from one piece of land to another, and can sometimes involve various contracting arrangements between the buyer and seller concerning the occupation of the land subject to the lease. Specialist advice and a carefully drafted agreement are vital in the latter area, as quota may move many hundreds of miles under these arrangements and the practical difficulties for the purchaser in personally occupying the leased land would be formidable.

All parties with an interest in land from which quota is being massaged must agree to the transfer. This might include the landlord of the farm from which quota is being transferred, who will have a particular interest in his or her own entitlement to the quota and may be concerned at the implications of the subletting used to effect the transfer. The operative date of the quota transfer is the date on which the lease commences and the transfer must be registered with the government's Rural Payments Agency.

'Land used for milk production'

It can now be seen that the nature of 'land used for milk production' is central to the ownership and transfer of milk quota. This important phrase was not defined in the original quota legislation, and it was left to the courts to interpret its meaning. The leading case is *Puncknowle Farms Ltd* v *Kane* [1985] 2 EGLR 8, decided in the Queen's Bench Division of the High Court. In the light of *Puncknowle*, areas used for milk production are:

- Forage areas used for grazing and conservation (of hay, silage etc) and arable areas used to grow corn and straw which is fed to the dairy herd.
- Buildings and yards which house or service the dairy herd.
- The dairy herd includes: dairy cows, dry cows, followers (heifers) intended as herd replacements and dairy bulls.

Disputes over areas used for milk production are settled by arbitration. Arrangements are set out in the Dairy Produce Quota Regulations, including a duty on the President of the RICS to appoint an arbitrator when requested to do so. The arbitrator's duty under the Quota Regulations is to base the apportionment on areas used for milk production in the last five years in which production took place.

An example may help to illustrate the apportionment.

Areas used for milk production: example of apportionment

Nathaniel has been the tenant of Blackthorn Farm since 1978 when he established his dairy farming business in his own name. In 1980 he bought the freehold interest in nearby grazing land at Whitethorn, where he keeps heifers and dry cows. In 1984 he was allocated 650,000 litres of quota and quota now stands at 800,000 litres due to subsequent purchases (and allowing for subsequent cuts in the original allocations of quota throughout the 1980s and early 1990s).

Areas used for milk production at Blackthorn are as follows:

- Yard and buildings: 0.5 ha
- Grazing land: 59.5 ha

Areas used for milk production at Whitethorn are as follows:

- Two fields totalling 20 ha

The total area used for milk production is therefore 80 ha and a very simple apportionment would equate to 10,000 litres per ha. Blackthorn would therefore carry 600,000 litres of quota and Whitethorn 200,000 litres. However, the landlord of Blackthorn may well take the view that the land there should attract proportionately more of the quota, because the main dairy herd is there as well as the milking parlour, cow housing and so on. The European Court of Justice case of *Posthumus* v *Oosterwoud* (Case C – 121/90)[1992] 2 CMLR 336 ruled out the notion of allowing for relative productivity in apportioning quota unless a Member State has laid down its own objective criteria for this purpose. However, this does not seem to preclude the use of concepts like the 'hectare month' to apportion quota. In practice this would involve analysing areas used for milk production, field by field, over the last five years of production and then apportioning the quota on the basis of 'hectare months'. Continuing with the example:

Blackthorn:

- Yard and buildings: 0.5 ha × 12 months × 5 years = 30 hectare months (hm)
- 20 ha grazing/silage × 10 months × 5 years = 1,000 hm
- 20 ha grazing/silage × 8 months × 5 years = 800 hm
- 9.5 ha grazing × 9 months × 3 years (most recent) = 256.5 hm
- 9.5 ha grazing × 7 months × 2 years (until three years ago) = 133 hm
- Total hectare months at Blackthorn: 2219.5 hm

Whitethorn:

- 20 ha × 7 months x 5 years = 700 ha months

Total hm on all areas used for milk production:
2,219.5 hm + 700 hm = 2,915.5 hm

Allocating quota between Blackthorn and Whitethorn on this basis:

Blackthorn:

$$800{,}000 \text{ litres} \times \frac{2{,}219.5 \ hm}{2{,}915.5 \ hm} = 608{,}000 \text{ litres of quota}$$

Whitethorn

$$800{,}000 \text{ litres} \times \frac{700 \ hm}{2{,}915.5 \ hm} = 192{,}000 \text{ litres of quota}$$

(Figures rounded for convenience)

On this basis, a little more quota is allocated to Blackthorn and a little less to Whitethorn. This illustrates the need for care on the part of Nathaniel if he were to transfer any quota away from the holding, to ensure that the areas used for milk production have been identified and recorded carefully.

Two other points need to be made in relation to 'areas used for milk production'. First the Rural Payments Agency undertakes spot checks as part of its surveillance of the transfer of milk quotas, and these will include any transfer involving more than 20,000 litres of milk quota per ha. Second, the agency itself can demand arbitration over the apportionment of areas used for milk production if it is not satisfied in a particular case.

Prospective Apportionment is a useful device when it is known, for example, that land is going to be sold in the near future. All the parties with a legal interest in the land and quota agree apportionment beforehand, according to areas used for milk production. The Rural Payments Agency is notified, and this apportionment then stays valid and binding for six months although it can be revoked within this time by the agreement of all the original parties.

At the time of its transfer, quota may be used or unused in that quota year. The purchaser of milk quota will also need to consider the effect on the butterfat base, as the butterfat bases of old and new quota will be weighted together to arrive at a new base.

Permanent transfer without land

Milk quota can be transferred without land when it is done 'to improve the structure of milk production at the level of the holding' (European Council Regulation). The transferor of the quota must be reducing or ending dairy production permanently, and the transferee must be increasing production (or intending to reduce his or her reliance on quota leasing). The purchaser must also demonstrate that he or she is engaged in milk production within six months of the transfer (although there are exceptions for natural disasters and the like). Prior approval from the Rural Payments Agency is required to transfer quota without land.

Temporary transfer of quota – quota leasing

Quota may be leased within the quota year between producers. These leasing arrangements are notified to the Rural Payments Agency and there are some restrictions on leasing. Used milk quota cannot be leased, quota which itself has been leased in cannot be leased out again, and quota which has been permanently transferred without land cannot be leased out within two quota years. Quota leasing has proved a useful measure for farmers who, having exceeded their own quota, fear the imposition of milk supplementary levy. It is also useful for farmers who currently hold quota but are unable to use it. Quota which is not used or leased during a quota year can be confiscated by the Rural Payments Agency (although there are detailed rules covering the circumstances for restoration).

The *Thomsen Case* (C–401/99) judgment by the European Court of Justice deals with non-producing quota-holders. These are people who own quota but have no cows and are not dairy farmers. The effect of the judgment is that quota holders who do not produce against their quota for one year will have it confiscated. Leasing will not prevent confiscation. Non-producing quota-holders have been given until 31 March 2004 to sell their quota or resume production. Confiscated quota may be restored if production is resumed within a certain period – currently six years, although this may be reduced to two years under current EC proposals. The Rural Payments Agency estimates that these provisions may affect over 7,000 quota holders and just over 10% of the national quota.

The government also intends to consult the agricultural industry on a proposed '70% usage rule'. This is already allowed for in EC legislation, but has not been implemented in the UK. If a producer uses less than 70% of their quota, all or part of the unused quota could be confiscated to the national reserve. Both of these recent developments have important implications for the market in leased quota, along with emerging proposals for the dairy industry in the forthcoming reforms of the Common Agricultural Policy.

The role of the rural payments agency

The Rural Payments Agency is the government agency in the UK which is responsible for administering all aspects of farm quotas and subsidies. It has regional offices throughout the UK, but the headquarters for milk quota is PO Box 277, Exeter, EX5 1WB. The Agency's website is at www.rpa.gov.uk. Milk Quota work was originally undertaken through the Milk Marketing Board, which was abolished in the early 1990s with the introduction of a free market in wholesale milk. The Intervention Board for Agricultural Produce then carried out the work before passing it to the Rural Payments Agency. Broadly, the agency deals with:

- Maintaining the register of quota holders.
- Maintaining the list of Approved Purchasers of wholesale milk (who carry onerous record-keeping obligations in relation to milk quota).
- Quota transfers.
- Calculation and collection of levies.
- Confiscation and restoration of milk quota (quota can also be confiscated if it has been obtained through fraud or on a failure to declare annual sales by a holder of direct sales quota, as well as for non-production in a quota year).
- Compensation payments for any further cuts in milk quota (farmers were compensated for several permanent cuts to quota in the 1980s).
- The production of wholesale milk production statistics on a month by month basis.

Each dairy farmer is allocated a Trader Registration Number ('TREG' number), which together with the DEFRA Holding Number for the particular farm, are important when communicating with the Agency. The Agency can provide quota records to farmers and others, and issues a number of deadlines for various transactions concerning milk quota. Examples include:

- 1 March: Notification of Permanent Transfer via a lease of land.
- 10 working days before 31 March: Permanent transfer without land.
- 31 March: Permanent transfer with land other than by lease, and temporary transfer of quota.

Implications of milk quota for property valuation

It should be clear that the availability of milk quota is a vital consideration when called upon to value a dairy farm. The prospective purchaser's bid for a dairy farm is bound to reflect his or her view on the adequacy of any quota offered with the farm. In the worst case of a dairy farm with no quota, a significant sum will be needed for the separate purchase of quota, whereas a dairy farm that includes a generous milk quota should see its availability reflected in its overall valuation. During the 2002/03 quota year, milk quota has not been trading at more than about 20 p/litre (and much lower than this for most of the year). Previous years have seen much higher prices, up to and including 30 or 40 p/litre. Although average milk yields are still no more than 4,500 to 5,000 litres per cow, some of our more efficient farmers are now regularly achieving 10,000 litres per cow. At this level of output, and given a typical stocking rate of two cows per ha, it is not difficult to see that the cost of quota per ha is of the order of £4,000. In other words, it can exceed the value of the land itself (and the cow for that matter!). Even at average levels of production, the value of quota involved in production is likely to be in the order of £2,000 per ha.

It is therefore important that the valuer should establish the position regarding milk quota. As well as the amount of quota, it is important to examine the history of milk production over the most recent five years of production in order to establish which areas have been used for milk production. Is there a need to apportion milk quota between different parcels of land for valuation purposes, along the lines of the last example? The existence of other parties with an interest in the milk quota and related land also needs to be considered. Other parties may include mortgagees, partners or shareholders in a farming business (especially where the land occupied by the business may be held under different arrangements from the business itself).

The position on tenanted farms raises particular questions. On the principle that quota is attached to land, the landlord has a strong interest in the quota allocated to a let dairy farm. Indeed on a tenancy granted since 1984 the landlord may have made the quota available to the tenant, and the tenancy agreement will probably set out the terms on which this was done. There should also be provisions for the return of the quota to the landlord at the end of the tenancy. Tenants who received the original allocations of quota in 1984 are entitled to compensation at the end of their tenancy for any quota which they have transferred to the holding, and for their share (if any) in what remains of the original allocation of quota (this is explained more fully at the end of the chapter). In all cases, the valuers will need to ensure that they understand the exact nature of the relationship between the landlord and tenant regarding both the current quota and the quota originally allocated to the holding. Rent reviews on tenancies regulated by the Agricultural Holdings Act 1986 (most of those which commenced before 1 September 1995, and a small number since) are required to disregard milk quota transferred to the holding by the tenant (section 15 of the Agriculture Act 1986). Most valuers therefore agree that other milk quota is a relevant consideration in a rent review. For tenancies covered by the Agricultural Tenancies Act 1995 (most tenancies granted since 1 September 1995) it is important to consider the terms of the tenancy agreement itself.

Particular types of valuation may also raise quota issues. The compulsory acquisition of land used for milk production is normally dealt with in a straightforward way which nevertheless is not easy to reconcile with the legal framework of milk quota. Negotiations between the acquiring authority and claimant need to establish whether the land to be acquired has been used for milk production in the last five years. Acquiring authorities would generally prefer to pay a lower price which does not reflect the inclusion of milk quota with the land they acquire. Therefore the quota is likely to be retained by the farmer (by agreement with the acquiring authority) who is losing part of his or her land on compulsory purchase, to be used on the remaining land. Straightforward so far, but if an arbitrator is called upon to apportion quota in the future the award may need to consider the land which has been expropriated. That part of the quota could theoretically be 'sterilised' as the farmer no longer has rights over the land on which it was used, and the acquiring authority is unlikely to be a dairy producer or quota holder. Legally, confiscation into the national reserve would seem to be the only way out. Practically, we should perhaps be grateful for our imperfect national record of land used for milk production as the problem outlined here is more theoretical than real!

Quota issues will also arise when valuing farms for tax purposes, Inheritance Tax in particular. Although milk quota is treated as a separate asset for tax purposes, the Capital Taxes Office of the Inland Revenue will accept its inclusion with the value of a dairy farm. In these circumstances, it will also accept that a claim for Agricultural Property Relief from Inheritance Tax will include the value of the quota with the agricultural value of the land. As for Capital Gains Tax (CGT), a gain on the disposal of quota is subject to CGT. This can be particularly painful for quota from the original allocation, as it has no base cost for CGT and therefore no basis for Indexation Allowance either. Quota which consists of both the original allocation and subsequently purchased quota will be 'pooled' in order to establish a base cost for CGT. Milk quota is however, eligible for 'Rollover' relief (Relief for the Replacement of Business Assets). Milk quota will generally be a business asset for CGT, thus qualifying for the more generous rate of Taper Relief. However, the farmer who gives up milking and then leases out quota for a year or more may find that it has been converted into a non-business asset. This was the issue in *Cottle* v *Coldicott* [1995] STC 239, where the taxpayer was denied CGT Retirement Relief on gains on milk quota because he had leased out the quota for a season after giving up dairy farming. Retirement Relief has now been phased out, but the distinction is still important for Taper Relief from CGT.

One of the most technically complex valuations arising in connection with milk quota concerns the calculation of end-of-tenancy compensation to a farmer who was originally allocated milk quota in 1984. The entitlement to compensation, and the valuation itself, is covered by the Agriculture Act 1986 (not, confusingly for the newcomer to this area, the Agricultural Holdings Act of the same year which covers most other aspects of the landlord–tenant relationship for agricultural tenancies granted before 1 September 1995).

End of tenancy compensation under the Agriculture Act 1986

A tenant is eligible for compensation for milk quota at the end of the tenancy if he or she is a tenant within the meaning of the Agricultural Holdings Act 1986 (in practice most agricultural tenants at that time), and

- Was in occupation on 2 April 1984.
- Had milk quota allocated in relation to land in the tenancy, or has acquired quota since then at his or her own cost.
- Alternatively, a statutory successor to such a tenant is eligible for compensation instead.
- The right to compensation may also pass to an assignee of either the original tenant or a statutory successor (in practice such assignments are rare on agricultural holdings). The Act also covers the position of sub-tenants (again rare in practice).

The principle of compensation can be expressed fairly simply:

- Each farm has a 'standard quota'. This is the amount of milk that it could produce in ordinary circumstances. For this purpose, farms are only differentiated according to whether they are disadvantaged, severely disadvantaged or on 'other' land. The designations of disadvantage refer to the classification of land by DEFRA (formerly MAFF) for the payment of hill livestock subsidies, and DEFRA keeps maps which show the various areas. Most dairy farms will be found on 'other' land. Farms are then differentiated according to the breed of cow they keep. There are three categories. The first is the Channel Island breeds (Guernseys and Jerseys), South Devons and 'breeds with similar characteristics'. The second category is Ayrshire and Dairy Shorthorns (and breeds with similar characteristics). The third category is 'Other', which therefore covers most of our national dairy herd – the Friesians and Holsteins. These categories are to be found in the Milk Quota (Calculation of Standard Quota) (Amendment) Order 1992, SI 1992 No. 1225. It can therefore be seen that most of our national herd is in the 'other' category of breed, on the 'other' category of land. The current order tells us that the standard quota for this combination is a standard quota of 7,140 litres per ha based on an average yield per ha of 9,000 litres.

- The actual quota on the farm is called 'relevant quota' and may exceed the standard quota. This may be for two reasons. The original award may have been more generous because the 1983 levels of production (or alternative reference year) on the farm were big enough to warrant a higher amount of quota. As well as this, or instead of it, the tenant may also have purchased further quota at their own expense.
- Quota purchased by the tenant is called Transferred Quota, and the tenant is eligible for compensation for all of this quota.
- Any quota in excess of the standard quota (excluding transferred quota) is called Excess Quota, and the tenant is entitled to compensation for all of this as well.
- Tenants are also entitled to compensation for a share of the standard quota if they have invested in the dairy fixtures on the holding.
- Thus a tenant may find himself entitled to compensation for transferred quota, excess quota and a portion of the standard quota. The calculation of this is best explained by an example.

Example: End of tenancy compensation: Blackthorn Farm

This example uses the facts of the previous example, in which 800,000 litres of quota was apportioned between the 'areas used for milk production' on Blackthorn Farm and land at Whitethorn Farm. Blackthorn Farm is the tenanted farm, and we now consider the assessment of quota for end-of-tenancy compensation purposes under the 1986 Agriculture Act.

Notes

Calculations

1. The first step is to establish the registered quota, which will be provided by the Rural Payments Agency. *Relevant* quota must then be determined, according to the method used in the previous example. Complications can arise if the registered quota is not in the name of the tenant, as the Act specifically refers to quota registered in the tenant's name. Relevant quota is 608,000 litres (previous example)

Relevant quota is 608,000 litres (previous example)

2. The next step is to apportion relevant quota between *Allocated Quota* and *Transferred Quota*. Allocated quota comes from the original award of quota to the farmer, and transferred quota is what he has subsequently purchased. In our example, 150,000 litres of the 800,000 litres of total quota has been purchased. These figures allow for subsequent cuts in milk quota, the calculation of which can add further complications to this part of the claim.

$608,000 \text{ litres} \times \dfrac{150,000\ l}{800,000\ l}$

= 114,000 litres of Transferred Quota, leaves: 494,000 litres of Allocated Quota The Transferred Quota will be eligible for compensation.

3. Now the 'relevant number of hectares' must be determined. This is likely to be less than the 'area used for milk production' in the previous example, because the definition in Schedule 1 of the Agriculture Act 1986 is not so broad ranging as that in the Dairy Produce Quota Regulations. It is confined to areas used for feeding and keeping dairy cows, and does not include areas used for heifer rearing or for growing corn to feed to cows for example. It is also necessary to determine land in these categories during the 'relevant period'; ie the period on which the farm's quota was based (1983 in most cases).

Although the 'area used for milk production' was 60 ha in the previous example, it emerges from a consideration of the stocking history of Blackthorn Farm that the 'relevant number of hectares' in the 'relevant period' is only 55 ha.

Notes   **Calculations**

4. Ascertain the Prescribed Quota, by reference to SI 1992 No 1225. This tells us that the prescribed quota for 'other breeds' on 'other land' is 7,140 litres. The SI also tells us that the 'prescribed average yield' is 9,000 litres, ie this is the amount which the land could reasonably be expected to enable a cow in one of the 'other breeds' to yield. The prescribed quota can be increased or decreased if the parties agree that the prescribed average yield should be changed for the land in question (or can persuade an arbitrator of this), due say to its inherently greater productivity. This is likely to be hard-fought in negotiations, and in many cases the standard figure will be used as the basis for compensation. *Grounds v Attorney-General of the Duchy of Lancaster* [1989] 1 EGLR 6 is the leading case in this area.

After hard negotiations, the parties agree that the reasonable yield to be expected would be 10,000 litres/ha. The prescribed average yield is therefore increased:

$$\frac{10,000\ l}{9,000\ l} \times 7,140 \text{ litres}$$

to give a prescribed quota of **7,933 litres per ha**

5. The **Standard Quota** for the farm can now be calculated. It is found by:
Relevant hectares × Prescribed Quota

55 ha × 7,933 litres/ha
= 436,315 litres

6. Determine **Excess Quota** by
Allocated Quota – Standard Quota
Excess Quota is eligible for compensation. Note that at this stage, the farm may be found to have less quota than its standard quota, or (unlikely) the same amount. Clearly where allocated quota is less than standard quota no claim for Excess Quota will arise.

Allocated	
Quota:	494,000
Less Standard	
Quota:	436,315
Excess Quota	
(litres):	**57,685**

7. Now consider the 'tenant's fraction' of the Standard Quota. The tenant's fraction determines the amount of standard quota for which the tenant will be compensated, and is conventionally expressed by the formula:

$$\frac{r}{r + R}$$

Analysis of the rent of Blackthorn during 1983 shows it to have been £125/ha (deducting for farmhouse, etc). Therefore, *R*: 55 ha × £125/ha = £6,875. Consideration of the written down value of

Notes

where r is the rental value of tenant's dairy improvements and fixed equipment (eg milking parlours, cowsheds, etc.) and R is the rent payable for land to accommodate the cows, buildings and so on. The valuation date (or period) for this is the 'relevant period' (as above – normally 1983), and apportionment may be necessary if there was a rent review during that year for example. Agreement on the fraction can be another contentious area in negotiations, not least because of the difficulty in establishing the rental value of tenant's improvements and fixtures.

Note that if the Allocated Quota is less than the Standard Quota, the Tenant's Fraction is reduced in proportion to the shortfall (1986 Act).

8. The various parts of the claim can now be summarised in order to arrive at the total amount of quota which is eligible for compensation.

9. The final step is to place a value on the quota which is eligible for compensation. Even this is not necessarily a simple matter. One approach is to take market values of quota on transfer, and to deduct allowances for marketing costs, the rental value of the land used to transfer quota, an element for 'anxiety' in years in which the supplementary levy is expected to arise. In other cases, attempts have been made to compare the value of land with and without quota.

Calculations

the tenant's improvements and equipment has led to an agreed rental value of £2,500 – r.
Tenant's Fraction:

$$\frac{£2,500}{£2,500 + £6,875} = 27\%$$

ie 27% of the *Standard Quota* is also eligible for compensation:
436,315 litres × 27% = **117,886 litres**

Transferred Quota:	114,000
Excess Quota:	57,685
Tenant's Fraction:	117,886
TOTAL (litres):	289,571

Say clean quota is currently trading at 20p/litre. Allowing for marketing costs, rental element, etc. leaves 15p/litre. Compensation therefore:
289,571 l × 15p = £43,436

Comment

This example started with the need to apportion milk quota between a tenanted farm and some other owner-occupied land. The entire holding to which the quota is attached is sometimes called the 'Euro-holding' in milk quota jargon. In simpler cases, this apportionment may not be necessary. There may also be some cases where the tenant has not acquired any further quota, so there will be no need to consider Transferred Quota either. The example has deliberately said little about the method by which a rental value is placed on tenant's dairy improvements and fixtures at the end of 1983, as this is an area where some artistry may still be required in negotiations. One approach may be to look at the written-down value of the items and to decapitalise this at an acceptable rate of interest. This and other aspects of the claim rely on good records being available. In their absence, the valuer's judgment based on his or her knowledge of local farming practices and values will be called to the fore. Some of the other uncertainties about these valuations have been noted already.

Finally

The object of this chapter has been to acquaint the general valuer with the practical importance of milk quota in the valuation of dairy farms, and to provide a basis for further study for valuers who wish to specialise further in this area. Valuers seeking further information might start with the following sources, as well as the regulations and cases already mentioned.

Recommended reading

Rural Payments Agency (2002) *A Guide to Milk Quotas*, Exeter, Rural Payments Agency

> This is the official guide to milk quotas, last updated in May 2002 and so currently the most up-to-date information on milk quotas

Edwards, J (1995) *Milk Quotas Explained* RICS Books

> Some of the information in this book is now a little dated, in particular some of the references to legislation. Nevertheless it provides a short readable account of milk quotas and very comprehensive information on all aspects of end of tenancy claims including a fully-worked example.

© Charles Cowap 2003

Chapter 2

Valuation of Woodlands

Introduction

Great Britain's 2.3 million ha of woodlands and forests (about 10% of its land area) are made up of a wide range of woodland types, ages and sizes. In turn there is a wide range of woodland owners, who have a wide range of management objectives, from optimising timber production and financial returns to the management for non-timber benefits such as wildlife, landscape and recreational enhancement. As a result, there is often more than one potential market or end use for forests and woodlands.

The growing of timber is a long-term investment, its economics – and hence its capital value – being dominated by the cost of time, and the associated uncertainties. The time intervals between planting the trees, thinning and (eventually) clear-felling are long. The production cycle or rotation will frequently take 50 years for a conifer plantation and can exceed 100 years for broadleaves. Thus, in effect, a crop of trees is grown without really knowing what demand there will be for the species, size and quality of the timber produced, due to the extensive timescale involved. Many would argue that it is difficult to predict with confidence the demand and hence potential value of most commodities within 1–5 years, let alone 50–100 years!

There are a wide range of factors that influence the market value of woodlands and as a result a wide variation in values is evident. These factors include location, site details and crop details in addition to market factors. Location and site details impact most on land values and crop details most upon timber values. Location can, as for all properties, greatly affect market values, and prices achieved for woodlands have varied considerably between different parts of the country. For large commercial coniferous woodlands, location is particularly important in relation to both the physical aspects of the site and the proximity to the timber markets. Aspects such as soil type, terrain, elevation exposure and the threat of windthrow hazard impact upon the growing potential of the tree crop. Access and proximity to the timber markets impact upon the existing or potential income generation opportunities from timber production. This in turn affects the land value. For smaller woodlands, with high amenity value or sporting potential, location is more important in relation to the wealth and interests of the local population, as the main value is likely to be generated from neighbouring landowners and local residents who are more interested in the amenity and sporting potential than future income generation for timber production.

Crop details – species, growth rate, timber quality and age – affect market values, especially as the age increases. Wide variations in values are evident when one makes comparisons: conifers with broadleaves; fast-growing species with slow-growing species; woods of high-quality timber with low-quality timber; young woods with mature woods. This variation in value is due to the fact that all these aspects impact upon the existing or potential income generation opportunities from timber production.

Market factors also impact on values. Prices for timber are not only affected by species, tree size, quality, ease of harvesting and extraction as mentioned above, but also by a range of different external factors. These include fiscal policy, through grant support and taxation, international timber prices, and exchange rates.

Properties containing volumes of marketable timber often require specialist survey and mensuration techniques. Having assessed the above factors, it should be possible to assess the market value by adding the value of the land and timber. It must be remembered, however, that potential buyers have varied motivations, circumstances and objectives, which may not be primarily financial. Thus care needs to be taken as small woodlands, in particular, have been sold for prices above the value of their timber and land, often due to their sporting potential or other inherent attractions. These other non-timber factors which influence the market are particularly difficult to assess.

Therefore different types of woodland require different valuation approaches. Furthermore the market for woodlands is relatively limited and fragmented and comparable data are difficult to acquire. Forestry valuations therefore pose particular problems to the valuer.

Woodland categories

In market terms there are essentially two categories of woodland: primarily commercial or investment woodland and primarily amenity, recreational or sporting woodland. Within these categories, there are many different woodland types in terms of species, age, structure and condition.

Primarily commercial or investment woodland

This category of woodlands tends to be comprised of large areas dominated by coniferous species, such as spruce, pine, larch and fir, often located in the uplands. The reason for this is that conifers grow considerably more quickly than most broadleaved species, and so have a shorter rotation length. The average growth rate of the existing state forests is close to $11m^3/ha/yr$ (ie between Yield Class[1] 10 and 12) although this has risen consistently as silvicultural techniques have improved. Furthermore many upland sites are infertile and exposed, so have limited alternative land uses. In turn they therefore have limited capital values, being significantly cheaper than more productive lowland areas. Indeed almost all of the afforestation over the last 80 years has been in the uplands and only conifers grow well on these 'poor' sites.

These woods tend to be managed with the main objective of maximising wood production, and in turn, profit. This does not, however, have to exclude other non-timber objectives such as amenity and recreation but the main value of these woodlands is related to their current and future timber value and the projected revenue stream from mature and non-mature trees. Notwithstanding the above, species such as oak and ash can produce more valuable timber than most conifers. Therefore some broadleaved and mixed woodlands may also fall into this category and should be valued accordingly. The key issue is the quantity, quality and value of the timber that the woods contain and its availability for harvest. Thus the current timber value and the potential timber value need to be carefully assessed.

Primarily amenity, recreational or sporting woodlands

This category of woodlands tends to be comprised of small blocks of broadleaved or mixed woodlands. Broadleaves tend to be slower growing than conifers and take longer to mature. The average

---

[1]   Yield Class is a measure of growth rate. The yield class can be derived from published Forestry Commission Yield Class curves.

growth rate is close to 5m³/ha/yr (ie between Yield Class 4 and 6) which is half the growing rate of conifers. Broadleaves also require more fertile soils and lower elevations, so tend to be concentrated in lowland areas which tend to have relatively higher site values. These woods are often managed for non-timber objectives such as sporting or amenity. A number of recent reports have indicated that much of this resource is under managed and that the timber contained in these woodlands is low quality and so low value. Thus many of these woodlands that are bought and sold have little real standing timber value but may have high amenity value, or recreational or sporting potential. The market value is therefore more a matter of judging potential demand for the property as a whole than of calculating the standing timber value or future returns from timber production. It is often the case that woods with high amenity or Ancient Woodland[2] status can often command a premium over prices fetched for 'commercial' woodlands. Thus, estimating their value is more generally determined by the evidence of recent similar transactions.

Farm woods

Woods may form an integral part of a farm or rural estate or they may represent a distinct entity in their own right. The majority of farm woods will be regarded as small (under 10 ha) and in many cases will be in the form of spinneys and shelterbelts, whose significance will be more as an enhancement to the farm as a whole, whether for amenity or shooting. Thus a valuer would need to assess whether the woods provide some advantage to the farm, such as shelter for livestock, providing a shoot or increasing the amenity value of the farm and thus making the property more attractive to a wider range of potential purchasers. Their value is thus difficult to quantify as the valuation cannot be related to the income generating potential of these, but to a more subjective assessment. The valuer needs to form an opinion on such considerations as the woodlands' contribution to amenity (which can vary according to location), the percentage of the farm that they comprise, and of course the likely potential purchasers, their goals

---

[2]   Ancient woodland: an area of woodland that has been continuously wooded since the 16th century and listed on English Nature's register.

and the sum they are willing to pay for the benefits of ownership. It should be noted that many of the areas planted with trees on farms were those which were never suited to agricultural cultivations and thus represent effectively an unproductive area. They may have been planted specifically to provide shelter from the wind or prevent soil erosion, and their removal may have unforeseen costs to livestock or crops, irrespective of timber value.

Forestry policy and market implications

Successive governments have sought to encourage both tree planting and sustainable management. Fiscal incentives, in the form of both grants and tax concessions, have been offered to promote these two aims.

A range of grants are currently available to encourage landowners to plant their land with trees. For farmland there are not only capital grants available, which may be sufficient to offset all of the planting costs, but also annual payments being made for up to 15 years to offset the loss of income from agricultural production. Notwithstanding these incentives, the relatively poor profitability of forestry and its long-term nature compared with agriculture has resulted in relatively small areas of land being converted to forestry over the last 20 years.

There are also a range of grants aimed at existing woodlands to help encourage sustainable management. These grants include payments to help offset the costs of undertaking works designed to improve the timber and/or the environmental attributes of the wood. Such work can include new access tracks, fencing, uneconomic thinning and scrub clearance as well as habitat management. Grants are also available for restocking a woodland (by planting or natural regeneration). However, these grants are less than half the grants available for new planting. Details of the rates of grant and the conditions applied to them can be obtained from the Forestry Commission.

Woodlands owners, both historically and currently, receive benefit from a range of tax concessions. Previously, it was possible to offset expenditure on forestry against other non-forestry income, while effectively enjoying freedom from taxation on the eventual sale of timber. These concessions were rescinded under the Finance Act 1988 and phased out by April 1993 with the result that 'commercial woodlands' were removed entirely from the scope of income tax and corporation tax. Thus, expenditure on the planting and maintenance of trees for timber production can no longer be allowed as a tax deduction against other income and proceeds from the sale of timber and capital planting grants are not chargeable to tax. (Growing timber is still exempted from capital gains tax and business property relief is available against inheritance tax.) As a result, there has since been a decline in the amount of expenditure on woodland together with a reduction in interest from private purchasers. Furthermore, the private purchaser interest has been shifting away from commercial woodlands to smaller amenity woodlands, so the sale of the larger commercial woodlands have become more dependent on investment from institutional and other funds. The personal tax position of prospective buyers can have a considerable bearing on the type of woodland they wish to buy and the price they are willing to pay.

While grants and tax concessions are offered as encouragement to plant trees and manage woodlands, there is also an element of enforcement on owners of existing woodland. No trees may be felled without permission from the Forestry Commission. This applies to any timber amounting to more than five cubic metres (equivalent to two large broad leaved trees), so that for even the smallest commercial operation such consent will be required. When there is a valid reason for the trees to be felled and removed, permission will, of course, be granted. However, this will normally be subject to a requirement that the land be replanted with an agreed species or mix of trees. Thus, when valuing woodland, it will usually be necessary to assume that the property will remain as woodland in perpetuity and that, under current forest regulations, there is no likelihood that it might be converted to farmland after the crop has been felled.

The property inspection

Preparatory work

Before inspecting the property, it is important to consider as many details of the factors pertaining to the valuation as possible. In this way, likely problem areas can be highlighted before the site visit and, as a result, given more effective consideration on the ground. Thus details of all relevant designations, statutory and non-statutory, that affect the property need to be gained. These may include the existence of Sites of Special Scientific Interest (SSSIs), Areas of Outstanding Natural Beauty (AONBs), Environmentally

Sensitive Areas (ESAs), as well as particular woodland designations such as Ancient and Ancient Semi-Natural Woodlands and Tree Preservation Orders (TPOs). These 'environmental' designations may restrict value by limiting or preventing economic forest management. Notwithstanding this, woodlands subject to such designations – in particular SSSIs and Ancient Woodland – can, however, often command premiums due to their perceived habitat and amenity status. Similar parallels can be drawn with buildings that are listed, which may attract premium values for historical or architectural reasons.

Access rights and other features of particular historical or cultural significance also need to be carefully assessed for their impact on value prior to the inspection. These include rights of access (to third parties or the public in general) granted by wayleaves, easements, and rights of way, together with liabilities for the maintenance of other features such as fencing (and other boundaries) and Scheduled Ancient Monuments. These aspects can also impact upon current and future timber operations as well as potential demand from prospective buyers.

Finally it is important to be aware of any 'non-woodland development' such as dwellings, buildings, plant, machinery and other equipment, materials and prepared products. There may also be an estate sawmill, preservation plant, and a forest nursery. If there is scope for future development potential, such as building development or mineral extraction, this must be carefully considered, and the relevant local planning controls need to be investigated.

The site inspection

Surveys are necessary not only to check or verify the legal boundaries, access routes, location of wayleaves and rights of way, but also to gain physical site details and crop details which are necessary to assess the current or potential timber value of the wood as well as its potential market. Using a 1:2,500 or 1:10,000 plan, the external boundaries, wayleaves, rights of way and fencing and maintenance liabilities can be verified, and the location of the access routes and internal tracks plotted. In addition, the wood may need to be divided into compartments and sub-compartments to reflect different crop types, in terms of age, species, composition and condition. The areas for each compartment and sub-compartment need to be ascertained and recorded (if not detailed on the map) and separately assessed. It is also necessary to determine whether these compartments are fully stocked. Typically 5–20% of woodland will be unstocked due to areas being used for roads, rides and streams as well as gaps resulting from crop failures.

Physical site details

Physical site details which should be appraised include:

1. Soil (type, depth, fertility, drainage and pH), climate (including temperature, wind and rainfall), and topography (aspect, altitude, elevation and slope) as these factors will influence the range of tree species that can be grown successfully together with their potential growth rates.
2. Crop stability, which can impact upon the rotation length. This can be quantified by subsequently estimating the Windthrow Hazard Class (WHC) with the use of relevant tables published by the Forestry Commission. These tables classify the average heights at which the onset of windthrow can be expected.[3]
3. The existence and diversity of ground flora, as this gives a guide not only to the soil type, but also the potential habitat value of the wood.
4. The presence of non-woodland features such as streams and ponds, ancient hedges and banks and archaeological features, as they contribute to the amenity and habitat value of the wood.
5. The access and extraction infrastructure and opportunities as these will impact on the timber harvesting costs, which will in turn affect the current standing timber prices.

---

[3]   Wind damage can be a serious threat in terms of windthrow, particularly in the more exposed upland regions. Besides windthrow, the risks include wind break and wind distortion, as well as retarded growth. The likelihood of windthrow is influenced not only by the windiness of the region, but also by the elevation, topography and soil conditions of the site. Taking these factors into account, the wind damage susceptibility can be quantified by referring to the Windthrow Hazard Classification booklet published by the Forestry Commission. Where the likelihood of windthrow is deemed to be high (ie those sites with a high Windthrow Hazard Class), it may be necessary to prematurely fell the crop, as the taller the crop becomes, the greater the chance of windthrow.

Crop details

Having appraised the site details, the crop details need to be considered. These should include:

1. The species, origin, pure or mixed, and proportion of mixture (area occupied by the canopy, not number of stems) which impact not only upon the potential rotation length, but also timber uses and market and hence timber values.
2. The silvicultural system and forest type (broadleaved, mixed, coniferous, coppice with standards, scrub and felled areas).
3. The age[4] of the trees (by means of records, whorls of branches or height and original spacing).
4. The current stocking density, tree diameters[5], basal areas[6] and

---

[4] **How to estimate the age of trees:** in the absence of planting records, the most accurate way of estimating the age of a crop of trees is by felling one or two to count the annual rings. Where felling is inappropriate, a rough idea of the age (primarily for conifers) can be gained by counting the whorls. Each new set of branches (a whorl) grows radially around the stem.

    Another method involves the removal of a core of wood from the stem of a tree using a hand-held hollow boring instrument (Pressler's borer). The number of annual rings and their distance apart will be seen on the core, allowing a rough estimate of age to be deduced. (Some experience is necessary to gain reliable estimates by this method.)

[5] **Tree diameters:** tree diameter is typically measured 1.3m above ground level. This measure is known as diameter at breast height (d.b.h.). A special timber girthing tape calibrated in centimetres (3.14 cm or π cm) is carefully placed around the circumference of the tree at breast height and the diameter is directly read off. (Trees with a d.b.h. of less than 7cm are assumed to have no volume and so are conventionally classified as unmeasurable.) More expensive digital instruments are also available.

[6] **Basal area:** the basal area of an individual tree is the cross-sectional area in m^2 of the tree at its breast height point. The basal area can thus be calculated by converting the d.b.h. into basal area (formula πr^2). Alternatively, conversion of d.b.h. to basal area is possible using published tables. The basal area of the crop can also be estimated using a relascope (a glass or plastic wedge prism relascope is the type in current use).

top heights[7]. This information can then be used to calculate the volume of the standing timber. The yield class of each species can also be subsequently ascertained (found by age and top height and the appropriate published yield class curves) which give a guide to current growth rates as well as future thinning and felling yields and intended rotation length.

5.   The quality and condition of the crop. Factors such as damage to the crop (in terms of the presence and extent of insect and animal pest damage, fungal decay, wind and fire damage) need to be assessed. The timber quality of the crop (in terms of straightness, form, presence of branching) also needs to be assessed along with areas of the crop that are 'in check' or unproductive.

6.   Past treatments including brashing, pruning and thinning.

This inspection of the crop will enable not only the landscape and habitat values to be determined, but also present content, volume and condition of the timber along with reliable estimates for future growth, thinning and felling proposals and, in turn, future income. The reliability of these forecasts depends upon the accuracy of the growing stock data and the thinning and felling policy being carried out as planned.

Calculating current standing timber value

The current standing timber value is calculated by assessing the volume of standing timber and then multiplying the volume by the relevant standing timber price. Thus it is necessary to first estimate the quantities of standing timber.

Estimating quantities of standing saleable timber

The valuer should take particular care if there are quantities of standing saleable timber as they are germane to the valuation.

There are several different methods of estimating the volume of

---

[7]   **Top heights**: height measurements must rely on geometrical or trigonometrical principles (apart from direct measurements involving climbing or felling trees!). The main instruments used are hypsometers or clinometers, though more expensive digital instruments are now available. (While a rough estimate of height is possible with basic instruments, such as a ruler or a straight rod, greater accuracy calls for the aforementioned specialist instruments.)

standing timber. The choice of method may be influenced by the size and quantity of produce, its value, and possibly by the local conditions and facilities available. This is an exercise that requires considerable knowledge in practice.

Forest measurement procedures are comprehensively described by Hamilton (1988) and Edwards (1981). These two sources are the basis of this section. It should be noted, however, that to gain an accurate estimate of standing volume involves quite complex methods of mensuration that are beyond the scope of this book. Thus only a brief guide is given below.

The volume of a standing tree can be derived from the d.b.h. (diameter at breast height) and the height of the tree. If a tree resembled a perfect cylinder, the volume could be derived from converting the d.b.h. to basal area and multiplying by the height (formula $\pi r^2 \times h$). However, the diameter of a tree narrows as the height increases (referred to as 'taper' which is conventionally expressed as a height to diameter ratio), thus resembling more closely the mathematical shape of a cone to reflect the fact that trees taper. Form height tables have been derived (Hamilton, 1988) which are an essential reference as this taper varies among tree species. An approximate estimate of this volume can be calculated by multiplying the total tree height by the basal area and then by the relevant form factor to give a total volume in m³. (The form factor varies between approximately 0.35 and 0.5 according to species, age and growing conditions.)

To estimate the standing volume of an entire crop, measure the d.b.h. of each tree within a sample plot, convert the d.b.h. of each tree to the basal area, add all these basal areas together and then multiply by the form factor to calculate the volume for the sample plot. This in turn can be converted to a per hectare or whole crop basis. High value crops require greater accuracy and therefore the volume of each tree may need to be ascertained.

An example of the basal area/form height method is shown below (Figure 1).

Standing timber prices

Prices for standing timber are extremely variable. The main factors which affect value are: species, tree size, quality, quantity being sold, ease of harvesting, access by road haulage and geographical location (proximity to the timber market). The main categories of timber for sale are sawlogs, and small diameter roundwood. Other

Figure 1 Volume Assessment of Standing Trees (Scots Pine)

Sample plot = 100m²

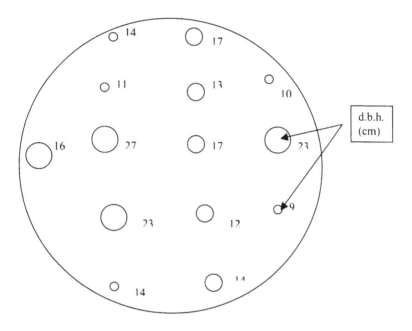

d.b.h. (cm)	No.	Basal Area (m²)		
9	1	0.0064		
10	1	0.0079	Total Plot Basal Area (m²)	0.3016
11	1	0.0095	Basal Area/ha (m²)	30.16
12	1	0.0113	Top Height (m)	16
13	1	0.0133	Form height (m)	6.83
14	3	0.0462	Plot Volume (m³)	2.06
16	1	0.0200	Vol/ha (m³)	206
17	2	0.0460		
23	2	0.0840		
27	1	0.0570		

Notes:
Top height gained by measuring with a clinometer the top height of the tree with the largest d.b.h. (ie 27cm) in the sample plot.
Form height derived from the Forest Mensuration Handbook.
Plot volume = form height × basal area.

categories include veneer logs (broadleaves only), pit props, transmission poles, fencing, turnery and minor products such as bark, chips and foliage.

There is a lack of published data on timber prices, particularly for broadleaved timber. Average prices for standing coniferous timber sold from Forestry Commission areas are published. The prices are reported in forestry magazines such as Timber Grower and Forestry and British Timber, etc. The data from these published sales (of coniferous timber) show that there is a good relationship between tree size and timber prices (regardless of species). Thus the larger the mean tree size, the greater the price per m³, within certain limits. Not only is the smaller timber (eg earlier thinnings) sold to the lower value small roundwood category, but the harvesting cost of smaller timber on a per m³ basis is greater. Thus making the relevant adjustments in terms of species, quality, access and location, the standing price can be assessed. While the same principles hold good for broadleaved timber, it should be noted that published data for broadleaves is more sporadic and less complete. This is due to the fact the broadleaved timber trade is more complex, with much greater variation in prices, depending on species, size, form, quality and marketing expertise of the seller.

Valuation methods

Forestry valuations, like all valuations, are required for a range of purposes, including sales and purchases, capital taxation, compensation and insurance. The purpose and basis of the valuation required clearly impacts upon the valuation method. This section focuses on open market valuations, though principles are similar for other types of valuation. A number of methods are used in forestry valuation to arrive at the open market value of the timber. These include the use of comparables, present market value and expectation value. All of these have a part to play in enabling the valuer to come to a considered opinion.

Comparables/market evidence

In certain cases, it may be possible to compare the woodland being valued to other properties that have recently been sold or are currently being offered for sale at specific asking prices. It should be noted, however, that woodland properties, like all properties, reach the market for many reasons and by various methods of sale.

They also vary considerably according to location, extent, age, structure and condition. Property prices also vary according to not only the economic, financial and political climate, but also due to fiscal policies, in particular the grants and tax concessions. In addition to these numerous variables is the fact there is a shortage of market evidence. Not only are there limited transactions, but also woodland sales tend to be handled by a small number of specialist agents, be they chartered surveyors or forest management companies. Results of these transactions are rarely disclosed. Price data, of course, can be gained by obtaining property particulars and bulletins from these specialist agents, and thus guide prices (subject to the usual caveats) can be estimated for different, types, ages and sizes of woodland property in various locations. However it is difficult to compare like with like – so varied are the location, extent and quality of sites and crops.

Notwithstanding the above-mentioned difficulties, it is often possible to use market comparables to gain an estimate of the value, particularly for smaller amenity woods, where the main market features may be ones of size and location. Comparables are likely to be of less help where the main market features are the volume and quality of the timber being grown, due to the greater number of technical variables. Thus it will often be necessary to consider separately the site or 'land value' and the standing timber value. The land will be assessed as 'planting land' and comparables used to assess its value (subject to making the necessary adjustments – in particular the physical site details and infrastructure).

The method of valuing the growing timber will then depend on the age of the trees at the time. When the trees are mature, or nearing maturity (ie their optimum rotation length), the present market value can be used to assess the current value of the standing timber. This figure is then added to the `land value' to help assess the market value of the whole property.

Where the trees are still young (less than 20 years), there is likely to be no meaningful timber value. Therefore assessing the present value of the timber is not appropriate. Instead it may be more appropriate to ascertain the net expenditure (ie the total establishment costs less government grants) incurred to date, and then add compound interest to the net costs incurred. However, past costs (often referred to as historic costs) are not a guide to future income generation and thus such a method needs to be treated with caution. Alternatively an 'expectation value' can be assessed to take into account future income and expenditure. This method is still problematic for young plantations but is considered more appropriate for semi-mature woods (as discussed below).

Present market value

The present market value approach is to assess the current standing timber value; ie the market value of the timber if cut now. It is often referred to as the 'devastation value'. The present market value can be appropriately applied to mature commercial woodland categories. However, the present market value approach needs to be treated with caution, as the value derived from this method can often produce a value that is less than the open market value, and for some woodlands can actually result in a negative value. For example a woodland that contains small and immature trees and/or the timber is of a poor form can cost more to harvest than the proceeds from the sale of timber. Yet comparable transactions clearly indicate that such woodlands have a market value over and above the value of just the planting land. This is because the present market value approach ignores both the non-timber value and the future timber value.

The current standing timber value is calculated by assessing the volume of the standing timber and then multiplying this volume by the standing timber prices, reflecting factors such as species, average tree volume, as well as location and access. The volume of the standing timber needs to be assessed for each woodland category according to species and age, as the species and age of the crop in turn have an impact on the standing timber value.

Expectation value

Where a crop is not yet mature, the timber may not have reached its full potential in terms of its size and value. Thus using a present value approach which estimates the current standing timber value will ignore this future potential value and thus may result in an undervaluation of the asset. Thus expectation value is often used for open market valuations as this method takes into account the future income and expenditure, which is likely to be more appropriate for such woodlands.

To assess expectation value, it is still, of course, necessary to fully inspect the crop to determine the relevant factors, such as the site details, species, age, stocking density, d.b.h., top height, yield class, volume and timber quality etc. Unlike the present value, however,

which estimates the current standing timber value, only future
income and expenditure are to be considered. Therefore it is
necessary to forecast the volume yield from thinnings and final
crop to the anticipated optimum rotation age, as these are the main
management operations that will result in future income and
expenditure. The Forestry Management Tables (Forestry Commission Booklet 48) can be used to determine the optimum economic
rotation length, and thus the expected volume yield from thinnings
and final felling to that rotation age. Using the above data from the
crop inspection an appropriate 'yield model' from Forestry
Commission Booklet 48 can be selected which sets out the future
trends of growth to estimate the probable development of that
woodland. The expected volume yield from thinnings and final
felling can be read directly from that yield model. These yield
models are tabular presentations of stand growth and yield,
including data on predicted tree diameters, stocking density, and
average tree volume as shown in Table 1 below. Yield models have
been prepared for all the major 'commercial' tree species in Britain
and for a wide variety of treatments including a range of initial tree
planting densities and thinning regimes.

It is important to note that it is inevitable that an individual
woodland/woodland compartment will vary in one respect or
another from the yield model, be it the actual physical stand
characteristics or future management regimes, which will be
influenced by market factors. Therefore it is necessary to tailor the
predictions given in the yield models accordingly. Furthermore, it
is also important to note that yield models assume full stocking and
thus should be adjusted to allow for open spaces such as roads,
rides, streams, buildings and crop failures.

Once future volume yields from thinnings and final felling have
been estimated, unit values, based on standing timber prices,
should be applied to the volumes at relevant points in time and
discounted to the present, using an appropriate discount rate, and
constant costs and prices[8].

The discount rate will clearly have a significant impact on the
capital value derived – the greater the timescale involved, the
greater the impact. Determining the appropriate discount rate to

---

8   There are difficulties in determining future costs and prices: the
     assumption of constant costs/prices over a long time period limit the
     usefulness of expectation value.

use will be dictated by the market, but will probably be taken to approximate to the level of internal rate of return being sought by investors at the time. This in turn requires a full awareness of the prevailing trends in the forestry investment market. Notwithstanding that, the discount rate adopted will also vary according to the woodland category, its age, timber quality and location.

Reinstatement cost method

Woodlands up to 20 years old are not likely to contain any saleable timber. Therefore the present market value approach is not appropriate. Expectation value is also problematic for such woodlands as they are many years away from their optimum economic rotation length. The further ahead that costs and prices are predicted, the more inaccurate such projections are likely to be. The discount rate will also have a very significant effect on the capital value derived.

Thus an alternative approach is to consider the costs of establishment. The net costs of establishment can be derived from actual or standard published costs. In addition, the availability of grant aid needs to be taken into account to ascertain the net cost. This net establishment cost should then be compounded to reflect the opportunity cost on the money invested for that period. However, such a method has its limitations as, firstly, it does not take into account the future income potential and, secondly, it does not reflect the state of the market. Thus the analysis of comparables is vital for younger woods.

Valuations for insurance

The valuation of woods for insurance follow similar principles to those described for open market valuations, the main difference being the exclusion of the value of the site itself. For woodlands up to 20 years old, the insurance value is based on the cost of establishment (net of grant) plus compound interest to allow for the delay that the owner will incur by having to replant the crop. Thereafter the basis of valuation is either present market value or expectation value.

Due to the relative complexity of valuing woodlands, some insurance companies calculate their own standard set of figures for specified age groups of different types of trees in both lowland and upland areas.

Table 1 Yield Model: Sitka Spruce, YC14; Intermediate thinning; 2m spacing

MAINCROP after thinning							Yield from THINNINGS					CUMULATIVE PRODUCTION		MAI	Age
Age yrs	Top ht	Trees /ha	Mean d.b.h.	BA /ha	Mean vol	Vol /ha	Trees /ha	Mean d.b.h.	BA /ha	Mean vol	Vol /ha	BA /ha	Vol /ha	Vol /ha	yrs
18	7.3	2311	11	24	0.03	66	0	0	0	0	0	24	66	3.7	18
23	10.2	1351	15	24	0.07	90	895	12	11	0.05	49	35	139	6.0	23
28	13.0	951	19	28	0.14	133	400	15	7	0.12	49	46	231	8.3	28
33	15.7	732	23	31	0.26	188	220	19	6	0.22	49	56	335	10.2	33
38	18.2	595	27	35	0.41	246	137	22	5	0.36	49	64	442	11.6	38
43	20.4	496	31	37	0.60	300	99	25	5	0.50	49	72	545	12.7	43
48	22.4	422	34	38	0.82	345	74	28	5	0.66	49	78	639	13.3	48
53	24.1	374	37	40	1.04	388	48	31	4	0.85	41	83	723	13.6	53
58	25.5	341	39	41	1.25	426	34	33	3	1.05	35	87	796	13.7	58
63	26.7	316	41	42	1.45	457	25	36	2	1.25	31	91	858	13.6	63
68	27.7	297	43	43	1.63	484	20	37	2	1.40	27	94	912	13.4	68
73	28.6	281	45	44	1.81	508	16	39	2	1.51	25	96	961	13.2	73
78	29.4	267	46	44	1.98	529	13	41	2	1.66	22	98	1003	12.9	78

Source : Edwards and Christie (1981) Forestry Commission Booklet 48.

Glossary of Terms used in Table 1:

Age	The number of growing seasons that have elapsed since the trees were planted.
Top Ht	Top height: the average height of a number of 'top height trees' in a compartment, where a 'top height tree' is the tree of largest d.b.h. in a 0.01 ha sample plot.
Maincrop after Thinning	All the live trees left in the stand, at a given age, after any thinnings have been removed.
Yield from Thinnings	All the live trees removed in the thinning.
Trees/ha	The number of live trees per hectare in the compartment.
Mean d.b.h.	The mean diameter, in centimetres, of all live trees measured at 1.3m (diameter at breast height) above ground level.
BA/ha	Basal area, in square metres, per hectare. The sum of the overbark cross-sectional area of the stems of all live trees measured at 1.3m above ground level.
Mean vol.	The average volume, in cubic metres, of all live trees, including any with a d.b.h. of less than 7cm.
Vol/ha	The overbark volume, in cubic metres per hectare of the live trees. In conifers, all timber on the main stem which has an overbark diameter of 7cm or over is included. In broadleaves, the measurement is either to 7cm, or to the point at which no main stem is distinguishable, whichever comes first.
Cumulative Production	The main crop basal area or volume of the present and all previous thinnings.
MAI	The mean annual volume increment; i.e. the cumulative volume production to date, divided by the age.
Note:	All trees which die through natural mortality are excluded.

Study 1

Pole stage to mid-age crops (20–40 years old)

The main methods of valuation to assess the value of this category and age of woodlands are comparable sales, expectation value or present value.

The present value of the wood can be assessed by estimating the volume of timber and multiplying by the standing timber price. The woodland at this age, however, is only likely to contain small diameter timber of relatively low value, so such a valuation approach ignores the future potential value of the wood and thus undervalues this asset. Accordingly expectation value which takes into account future income and expenditure is likely to be more appropriate.

The application of expectation value and present value are shown in the example valuation, as follows.

10 hectares of Scots pine, planted 25 years ago. The crop is fully stocked, there is good access to the wood and there are no felling restrictions. The wood has been inspected and the following field data has been ascertained:

Notes/sources of information :

Age	25 years	Estate records/count annual rings.
Top height	12m	Top height of fattest girthed tree in sample plot.
Trees/ha	1,300	Counting trees in sample plot.
Mean d.b.h.	14cm	Mean of d.b.h.s in sample plot.
BA/ha	20m^2	Convert d.b.h. to BA, then multiply by trees/ha.
Mean vol.	0.07	Convert top height to form height, then
Vol/ha	91m^3	multiply by BA.
Yield Class	12	Estimated from published yield class curves, based on top height and age of the crop.
Rotation length	65	Estimated from published yield class curves.

Using these data, the present value of the standing timber crop can be assessed as follows :

Present Value

Scots pine	Per ha	Per site
Volume/ha	91m^3	
× Standing Value[1] @	£ 5/m^3	
Present Value of timber	£ 455	
Value of land[2]	£1,000	
Total Value	£1,455	
Total Value (10 ha)		£14,550

[1]   The average tree size was 0.07m^3, based on an average d.b.h. of 14 and top height of 12. Tree sizes of this volume are likely to be classified as small diameter

timber and thus sold to the pulp and chipboard industries. Using current
market price data for standing timber (taking into account species, timber
quality, access and location) the current market price was estimated to be £5/m³.
2   The value of the land is based on comparable sales of bare land which is
required to be restocked with trees, thus restricting the value of the land.
3   Timber prices used in Study 1 are based on prices in March 2001. These are
illustrative only. Prices have reduced since then.

Expectation Value

Scots pine	per ha	per site
Age	25 years	
Rotation length⁹	65 years	
Discount period	40 years	
Yield Class	12	
Final crop volume	367m³	
Standing value of final timber crop¹⁰	£25/m³	
Final crop value	£9,175	
Discounted for 40 years @ 5%¹¹	0.142	
Expectation value of timber	£1,300	
Value of land	£1,000	
Total Value	£2,300	
Total Value (10 ha)		£23,000

9   The Forestry Management Tables (Forestry Commission Booklet 48)
can be used to determine expected rotation length and the expected
volume yield from thinnings and final crop to that rotation age. Note,
however, that the management tables should only be used as a guide
and future forecasts should be adjusted to reflect the actual crop
details.   For the above example, income from thinnings has been
excluded as it is assumed (arbitrarily in this example) to equate to the
interim costs of management and maintenance etc.
10   The average tree size was forecasted to be 1.4m³, which was derived
from the above-mentioned Forestry Management Tables, based on a
rotation length of 65 years for Scots pine, Yield Class 12. For tree sizes
of this volume, a high percentage of the tree should be saleable for the
sawlog market, thus attracting higher prices. The price was estimated
to be £25/m³ based on current market price data for standing timber.
11   Discounting has been used to obtain the present value of the projected
income in the future. A discount rate of 5% has been used for this
example. The discount rate will depend on the location and quality
of the wood and the current market rate. The current market rate is
in the region of 3%–5%. The choice of discount rate will have a very
significant affect on values. If a discount rate of 3% had been used in
the above example the value would have increased from £1,300 to
£2,813 (excluding the land value).

Using the above data, the resultant values derived from these two valuation approaches vary very significantly. While the expectation method is considered more appropriate, the value produced is very dependent on the discount rate. Thus both full awareness of discount rates used and comparable transactions is clearly essential.

Recommended reading

Edwards, P. N. and Christie J.M. (1981) *Yield Models for Forest Management*. Forestry Commission Booklet 48. Edinburgh: Forestry Commission.

Hamilton, G.J. (1988) *Forest Mensuration Handbook*. Forestry Commission Booklet 39. Edinburgh: Forestry Commission.

Chapter 3

Valuations under s18(1) of the Landlord and Tenant Act 1927[1]

At common law the measure of damages for breach of a covenant to repair is the diminution in the value of the landlord's reversionary interest, and events after the determination of the lease do not affect the matter except that they may be evidence of what was in prospect at the time the lease came to an end.

In most cases the cost of repairs is the best possible guide to the assessment of damages, and this is especially the case where the landlord intends to do the work or where an incoming tenant would, of necessity, have to carry out the work before he could properly occupy the premises. In connection with this latter point, what would be bound to happen, and would be reasonably foreseeable, is that the incoming tenant would use the disrepair as a bargaining point and thereby obtain a reduction in rent, or a rent-free period, so that the value of the landlord's reversion will therefore clearly be diminished. Where the landlord does the work the court would normally accept that the cost thereof equates to diminution in the value of the landlord's reversion: see *Jones* v *Herxheimer* [1950] 2 KB 106.

In some cases it is obvious that repair without improvement is futile. This is the case where an incoming tenant would then carry out an improvement which would effectively strip out the repair. An example of this is where there is a very old style fluorescent light fitting which even in repair would be removed and replaced by a category II fitting. In these cases the repairs are superseded and thus rendered valueless, and clearly there would be no diminution in the reversion by virtue of such repairs: see *Mather* v *Barclays Bank plc* [1978] 2 EGLR 254.

---

[1]   This chapter is directed towards the consideration of the limit on damages for dilapidations. It does not extend to a consideration of the limit on damages for breaches of any other covenants where *Tito* v *Waddel* could in many cases be in conflict with *Joyner* v *Weeks*.

The question as to whether the disrepair will be rendered valueless is one which has to be answered as at the date when the covenant ought to have been performed and not solely as a result of utilising hindsight: see *Cunliffe* v *Goodman* [1950] 2 KB 237. In other words where the supersession is inevitable as a result of the settled intention of the landlord or by the inevitability of circumstances which existed as at that date, ie the need to modernise and improve the property in order to render it lettable.

All of the above matters were considered in the case of *Shortlands Investments Ltd* v *Cargill plc* [1995] 1 EGLR 51. Shortlands held an underlease of an entire office building comprising 150,000 sq ft at a rent of £3,396,000 pa, which was agreed on review in September 1991. They sublet three floors of the building to Cargill in January 1985 with the tenant having a break clause which could be effected in January 1991, the letting was on internal repairing terms. Cargill exercised the break clause and Shortlands relet the premises in October 1992 on the basis of paying a reverse premium of £690,000, being an estimate of the sum required to bring the premises up to the normally acceptable letting conditions, although not all of the work required was covered by the tenants liability (for example, there was no covenant to remove the tenants' partitioning). Shortlands claimed damages from Cargill based on a terminal schedule of dilapidations, claiming that the cost of carrying out the disputed repairs was the proper measure of damages. Because Shortlands had not carried out any repairs, Cargill submitted that Shortlands had not suffered any loss, therefore the cost of repairs was not the appropriate measure. Cargill argued that section 18(1) of the Landlord and Tenant Act 1927 limited the diminution in value of the reversionary interest because in the then current market conditions the incoming tenants required specification changes which would render unnecessary certain repairs. Judgment was given for Shortlands in the sum of £294,934.47 being the calculated diminution in the value of the landlord's reversionary interest. Events after the determination of the lease were not to affect the matter except that they may be evidence of what was in prospect at the time the lease came to an end.

What was of particular interest in this case was that the value of Shortlands interest was negative and that Cargill were only liable for some of the items of disrepair. The court decided that the only way of assessing that part of the difference in value of the reversion was by examining the cost of repairs and then applying the limits imposed by section 18 (1). In determining the diminution in value for the purposes of the first limb of section 18 (1) the test derived from *Cunliffe* v *Goodman* was appropriate; ie viewing the question as at the date when the covenant ought to have been performed, was it inevitable, either because of the settled intention of the landlords, or for some other extraneous reason, that the premises would be pulled down or altered in such a way as to cause diminution of the value of the reversion. On the evidence, what was bound to happen, and what was reasonably foreseeable, was that the incoming tenant would use the disrepair as a bargaining point so that the value of the plaintiffs reversion was diminished. In considering the application of section 18(1) to the gross cost of repairs and loss of rent, the plaintiffs' reversionary interest always would have had a negative value and one had to assume a transaction under which something was paid to the transferee of the interest. The diminution in the reversionary interest was the difference in the amount which would have been paid by the willing transferor of that interest to the willing transferee if the premises were delivered up in a condition in conformity with the covenants and the amount paid out by the willing transferor to the willing transferee if the premises were delivered up in their actual condition. Because the common law claim was greater than the diminution in value so assessed, judgment was for the latter.

At the time that Cargill exercised its option to break almost half of the building was empty. The exercise of the break option was no doubt connected with the fact that the commercial property market was collapsing badly and the market was described as being 'in free fall'. The position of Shortlands was as pig in the middle having to pay rent subject to an upwards-only revision and other outgoings with much reduced income coming in. To make matters worse, overbuilding in the 1980s resulted in there being much unoccupied new office space and Shortlands were competing for tenants with the owners of brand new buildings.

Cargill's premises had been fitted out to a good standard and there was a need to serve a schedule of dilapidations after Cargill had moved out, mainly in relation to decorations and repairs revealed by the removal of tenants fixtures and fittings. As already stated, Cargill were not obliged to remove the non-structural partition walls nor to replace carpets. Notwithstanding that a schedule of dilapidations had been served prior to the term date of the lease, Cargill did not attempt to do any of the work specified in the schedule and they vacated, after having removed certain items of equipment, on or before 29 September 1991. As at that date Shortlands had not made any decision as to what works, if any, they would carry out at the premises and they marketed them as they were.

A new letting of the premises was agreed by Shortlands in June 1992 on the basis that they would pay £690,000 plus VAT to the incoming tenants. This was their estimate of the sum required to bring the premises up to the normally accepted condition for the letting of such premises by a landlord. That work included the removal of the partitions and work consequential thereon. It also provided for new carpets throughout. It was however plain from the evidence given by Shortlands' surveyor that the negotiations with the incoming tenants was not done in the same way as the landlord's building surveyor might discuss figures with an outgoing tenant's building surveyor, but it was equally plain that the fact that the premises were in disrepair gave the incoming tenant a bargaining counter which enabled them to demand and obtain a sum specifically for the disrepair, in addition to the other payments and special terms which they were able to negotiate as incentives to take the lease. The landlord's surveyor made the following statement in evidence:

> I agree with the general principle that [in September 1991] landlords were offering very attractive incentives to incoming tenants. The vital question is: would a 'reasonable landlord/developer' offer an incoming tenant a sum of money to repair and redecorate the premises when they were in a good state of repair and decoration already? One has only to ask the question to answer it. If the condition of the premises was sub-standard because of failure to repair and decorate, logically the landlord would have had to offer bigger incentives than if the premises were in a good state of repair and decoration. No amount of mixing up of repair and decoration on the one hand and 'fitting out' on the other can defeat this logic.

This evidence was accepted by the court.

Three leases were subsequently granted by Shortlands each being for the residue of 25 years from 29 September 1991 at a rent of £302,000 pa. The incoming tenants carried out substantial works of refurbishment, including some of the remedial work which it was alleged that Cargill should have carried out, and the court accepted that no one in September 1991 could have expected to foresee that a future tenant would fit out the premises to the very high standard chosen by the incoming tenants. The end result would have been different had it been possible to make this kind of forecast, or to similarly forecast that the landlord himself would have had to carry out work to a similar standard. Cargill contended that the extensive and lavish refit rendered the claimed repairs unnecessary, ie that they were superseded in the way argued in the *Mather* case. This contention was not accepted by the court because of its non forseeability.

In its approach to the assessment of damages the Court first considered the common law measure as quoted in *Woodfall's Law of Landlord & Tenant* (1993) para 13.081 and p 13/60 as follows:

> At common Law the Measure of Damages for breach of covenant to leave in repair at the end of the term was the cost of putting the premises into the state in which the tenant ought to have left them.... In addition the landlord is entitled to damages of loss of rent during the period needed to carry out the repairs.

The period referred to as 'the period needed to carry out the repairs' is not in the writer's opinion merely the period for doing the work, but it includes the lead up period including the time necessary to prepare specifications, to go out to tender, etc.

The court also quoted from *McGregor on Damages* 15th ed, para 1001 to 1003 as follows:

> 1001 Where the action is commenced after the expiration or earlier determination of the term the damages at common law are such a sum as will put the premises into the state of repair in which the tenant was bound to leave them. The first clear judicial holding that this was the proper measure is that of Denman J in *Morgan* v *Hardy* followed by the Court of Appeal in the leading case of *Joyner* v *Weeks*. In this latter case the choice between the two measures of cost of repairs and diminution in the reversion's value was directly before the court, and it was decided that the former represented the true measure; whether or not it exceeded the latter was a question that need not be explored. Lord Esher MR said that to award the cost of repairs was such an inveterate practice as to amount to a rule of law, while Fry LJ regarded such a rule as one of great practical convenience since it was far simpler than the alternative one and he had no hesitation in endorsing it. Megarry V-C in *Tito* v *Waddell (No 2)* questioned whether this case did indeed lay down that the cost of repairs was the invariable rule of damages, but that this was the universal interpretation is clear from the decisions dealt with below and from the intervention of statue, otherwise uncalled for, in the form of Section 18(1) of the Landlord and Tenant Act 1927.
>
> 1002 The cost of repairs is the short way of expressing the normal measure; more precisely it should be expressed, as by Wright J

in *Joyner* v *Weeks,* as the cost of repairs with some allowance for loss of rent or occupation during repair and with some deduction, where proper, by reason of substitution of new for old...

1003 The cost of repairs would in the normal case properly represent the extent to which the lessor has been injured by the breach. But where the circumstances were such that the loss to the lessor was less by reason of the fact that the repairs were not to be done completely or so thoroughly or were to be done by a third party at no expense to the lessor, the normal measure still applied; all these factors were in effect held to be collateral and did not go in mitigation. Thus no reduction of the damages was made in all the following circumstances

(1) where the landlord had decided before the end of the lease to pull down the buildings and had done so before action: *Inderwick* v *Leech;* (2) where the lessor had made an agreement with a third party to grant him at the end of the defendant's term a new lease under which the buildings were to be pulled down by the third party, the conditions of the premises having formed, it was said, no ingredient in the price: *Rawlings* v *Morgan;* (3) where owing to a deterioration in the neighbourhood, the premises would command as high a rent even though the covenant to repair was not strictly complied with: *Morgan* v *Hardy* followed in *Anstruther-Gough-Calthorpe* v *McOscar;* (4) where the Plaintiff had granted a new lease to a third party to run from the expiration of the defendant's lease under which the third party covenanted to repair, so that the performance of the defendant's covenant to repair was a matter of pecuniary indifference to the plaintiff: *Joyner* v *Weeks.*

Shortlands relied heavily on *Joyner* v *Weeks* [1891] 2 QB 31 where Lord Esher said at p43:

for a very long time there has been a constant practice as to the measure of damages in such cases. Such an inveterate practice amounts, in my opinion, to a rule of law. That rule is that, when there is a lease with a covenant to leave the premises in repair at the end of the term, and such covenant is broken, the lessee must pay what the lessor proves to be a reasonable and proper amount for putting the premises into the state of repair in which they ought to have been left. It is not necessary in this case to say that it is an absolute rule applicable under all circumstances; but I confess that I strongly incline to think it is so. It is a highly convenient rule...it is...I think, at all events, the ordinary rule...

In *Joyner* v *Weeks* the court disregarded the fact that before the lease expired the landlord had entered into a contract with a new tenant under which the tenant was to pull down part of the premises and repair the rest. Counsel for the defendant in the Shortlands case invited the court to hold that the *Joyner* v *Weeks* principal was not an invariable rule of law and that it did not apply in this case. To this extent they relied upon *Tito* v *Waddel (No 2)* [1977] Ch 106 to the extent that it was fundamental to all questions of damages that they are to compensate the plaintiff for his loss or injury by putting him in the same position as he would have been in had he not suffered the wrong. The question is not one of making the defendant disclose what he has saved by committing the wrong, but one of compensating the plaintiff. Therefore you are not to enrich the party aggrieved; you are not to impoverish him; you are so far as money can, to leave him in the same position as before.

The reason for the argument as to whether *Joyner* v *Weeks* applied was to establish the principle that there was a common law measure of damages by reference to the cost of carrying out the work. If *Tito* v *Waddel* applied then the entire argument would be based on the loss caused by the failure to carry out work to Shortlands. The court considered that if Shortlands had suffered no damage, or only minimal damage, then one does not begin to assess damages but in the light of the evidence given to the court it was plain that Shortlands had suffered damage because if the premises had been delivered up in a state of good repair and decoration they would have had a lesser negative value than they had as a result of the condition in which they had actually been delivered up. That worse negative value was evidenced by subsequent events and Shortlands had to offer a very large sum of money specifically related to the condition of the premises. Therefore the court was of the opinion that the rule in *Joyner* v *Weeks* stands today subject only to section 18(1) of the Landlord and Tenant Act 1927 and therefore 'because the measure of damage is the difference in value of the reversion at the end of the lease between the premises in their then state of unrepair and in the state in which they would have been if the covenant had been fulfilled. Matters happening after the lease came to an end do not affect the matter except that they may be evidence of what was in prospect at the time the lease came to an end. In most case the cost of repairs is a good guide to the difference in value of the reversion'.

Section 18 (1) imposes a limit on the recoverable damages in the following terms:

> Damages for breach of a covenant... to keep or put premises in repair during the currency of a lease, or to leave or put premises in repair at

the termination of a lease...shall in no case exceed the amount (if any)
by which the value of the reversion (whether immediate or not) in the
premises is diminished owing to the breach of such covenant... and in
particular no damages shall be recovered for a breach of any such
covenant...to leave or put premises in repair at the termination of a
lease, if it is shown that the premises, in whatever state of repair they
might be, would at or shortly after the termination of the tenancy have
been or be pulled down, or such structural alterations made therein as
would render valueless the repairs covered by the covenant.

The words in section 18 after the words 'and in particular' on their
literal meaning was not relevant to this case. There was no
demolition and there are no structural alterations. However, since
those words are plainly intended to be by way of example, they may
be some guide to the meaning of the earlier part of the section. Thus
the reading of the second part does not rule out a consideration of
non-structural alterations in the first part of the section, if non
structural alterations will cause material diminution to the value of
the reversion.

On the *Joyner* v *Weeks* basis the common law claim was
determined by the court as follows:

	£
Cost of works	89,381.87
Loss of 11 weeks rent	183,375.90
Loss of 11 weeks service charge	54,404.92
11 weeks void rates	28,179.07
	355,346.76

In so far as considering section 18 (1), the valuation of the leasehold
reversion was entirely hypothetical. The landlord's interest actually
extended to the whole of the building but the valuation was only
concerned with that part which had been leased by the defendants.
The valuations were also made on the basis that the leasehold
reversion was to be sold to an investor and in practice it would not
be so, as in law it could not be so because of the covenant in the lease
which forbade it as it would have been an assignment of part. In a
normal case where a property is transferred from one person to
another it is for the transferee to pay money to the transferor. In this
case the transferor had to pay money to the transferee to persuade
him to take on the burdens of the property transferred. It was pure
semantics as to which party was considered to be the willing
purchaser and which the willing seller, for what was relevant was
that the diminution in value was the difference between the amount

of money paid out by the willing transferor to the willing transferee, if the premises were delivered up in a condition in conformity with the covenants, and the amounts paid out by a willing transferor to the willing transferee if the premises were delivered up in their actual condition. In both cases there would be no willing purchaser if no money was paid to them. The principal difficulty which Cargill's had with this concept was that they could not accept that once the value of a property gets down to 'nil' there cannot be any diminution in value. That may be true in many, or perhaps most cases of chattels. It is quite a different situation where an owner of a leasehold property has an onerous interest which he wishes to transfer. Such an interest is transferable on the market if not 'saleable'. If one assumes a willing transferor and a willing transferee there will be a point in negotiations for a payment from the transferor where the parties are willing to do a deal.

The approach to the valuation was accepted as being 'the residual approach'. Therefore two valuations were prepared showing the two separate scenarios. The first with the property in the state as required by the lease and the second being for the property in its actual condition.

Cargill challenged some of the actual figures in both calculations in addition to challenging the cost of the works. The court held that:

> ...those challenges are of no importance, because the figures are reproduced in each schedule and since the only important result is the difference between the end result of each schedule, the size of figures in common between them is of no importance so long as they remain common. For example, under the heading of 'Deductions', schedule A begins with an item of 'Deficit of headrent for 30 weeks'. That period is made up of four weeks to remove partitions and lay carpets and 26 weeks for marketing period. None of that period of 30 weeks is for the defendant's account: it is a period in relation to which the plaintiffs would have suffered expenses even if the defendants had complied with their covenants and is compared with a period of 41 weeks' deficit of headrent incurred in the actual state of repair and decoration. There had been some discussion that a marketing period of 52 weeks would have been more appropriate. It makes no difference and I make no finding about it. If the marketing period should be 52 weeks, for the period of 30 weeks we would substitute 56 weeks and for the period of 41 weeks we would substitute 67 weeks and the difference would be the same, namely 11 weeks, which would be translated into cash sums each with similar differences.

The two calculations provided to the court were as follows:

A. Valuation assuming Tenant's compliance with covenants
A.1 Estimated value let and in good repair

	36,888ft²	at	£23.50	866,868
Less Headrent:	36,888ft²	at	£17.62	649,967

Profit rent	216,901
89 years' purchase at 17% – sinking fund attracting	
4% and 35% tax thereon	5.82
Estimated value let and in good repair	1,262,364

A.2 Deductions: £

(i)	Deficit of headrent for 30 weeks	374,981
(ii)	Non-recoverable service charge for 30 weeks	148,390
(iii)	Void rates	55,626
(iv)	Interest on rent, service charge and rates (see statement)	27,079
(v)	Letting fees (15% + VAT)	152,785
(vi)	Marketing costs	50,000
(vii)	Profit to purchaser @ 25% of gross value above	315,591

(viii)	Works a. Cost	165,648	
	b. Fees and VAT	48,452	
	c. Interest thereon for 6 months @ 12% pa	12,846	
	Total cost of work		226,946
(ix)	Acquisition costs say		10,000
	Total deductions unrelated to purchase price		1,361.398
			−99,034
(x)	Interest for 30 weeks on purchase price at 12% (credit)		−6,412

A.3 Net value: estimated value unless and in state of repair envisaged by tenants' covenants	−92.622

B. Valuation in actual state of repair and decoration
B.I Estimated value and let and in good repair

	36,888ft²	at	£23.50	866,868
Less Headrent:	36,888ft²	at	£17.62	649,967

Profit rent	216,901
89 years' purchase at 17% – sinking fund attracting 4% and 35% tax thereon	5.82
Estimated value let and in good repair	1,262,364

BA.2 Deductions:			£
(i)	Deficit of headrent for 41 weeks		512,474
(ii)	Non-recoverable service charge for 41 weeks		202,800
(iii)	Void rates		83,805
(iv)	Interest on rent, service charge and rates (see statement)		48,033
(v)	Letting fees (15% + VAT)		152,785
(vi)	Marketing costs		50,000
(vii)	Profit to purchaser @ 25% of gross value above		315,591
(viii)	Works a. Cost	235,464	
	b. Fees and VAT	68,873	
	c. Interest thereon for 6 months @ 12% pa	18,260	
	Total cost of work		322,597
(ix)	Acquisition costs say		10,000
	Total deductions unrelated to purchase price		1,698,085
			–435,721
(x)	Interest for 41 weeks on purchase price at 12% (credit)		–38,500

A.3 Net value: estimated value unless and in state of repair envisaged by tenants' covenants		–397,221

The court adjusted the net value in disrepair (–£397,221) by adding back a total of £9,277.63 being in respect of a disputed item of disrepair so that it amended the value in disrepair to minus £387,943.37. It therefore calculated the diminution in the value of the reversion as follows:

	£
Value in disrepair	–387,943.37
Value in repair	–92,622.00
Diminution	–295,321.47

A detailed examination of the input figures of the calculation shows that in calculating the value in repair regard was had to the period which it would take to let the property, the losses of service charge and rates which would occur during this letting period, and interest thereon. It also had regard to letting fees, marketing costs and a profit to the purchaser. This profit to the purchaser was required as he would take the risk of finding a tenant within the specified period at the specified rent. There was further deducted the necessary costs to make the property suitable for letting and the associated costs and interest thereon. The final item was an interest credit because the transferee would have the benefit of the cash sum of £99,034 for a period of 30 weeks and therefore would earn interest on that sum.

A consideration of the value in disrepair merely shows the same items, save that the item for the loss of rent during the letting period is increased to reflect the added period of time necessary to do the work, which in this case was taken at 11 weeks. This therefore had a knock on effect on the normal recoverable service charge and rates and therefore also on the interest thereon. The letting fees, marketing costs and profit were all the same, as the latter was based on a percentage of the gross value. The cost of the works was different because these reflected the expenditure on repairs. Finally the credit interest was different because the payover to a transferee was larger and held for a longer period of time.

The diminution in the value of reversion of £295,321.37 compares with the *Joyner* v *Weeks* figure of £355,346.76 and consequently by virtue of section 18(1) damages were limited to the former. It is interesting to see from this decision that the case of *Joyner* v *Weeks* was still very much in evidence and it was further noted that section 18(1) only relates to damages in respect of a breach of a covenant to repair. Where there is a breach of a covenant which is not a covenant to repair, for example a covenant to redecorate in the last year of the term or a covenant to remove partitioning, then section 18(1) will not apply and the rule in *Joyner* v *Weeks* will then apply subject to the argument raised in *Tito* v *Waddel*. The court in the *Cargill* case did not consider the case of *Ruxley Electronics & Construction Ltd* v *Forsyth* [1994] 1 WLR 650 which supports the proposition that the choice between damages assessed by reference to diminution in value and damages assessed by reference to the cost of works depends in part on whether the result of the breach has been to damage something which cannot readily be replaced. In such a case the plaintiff cannot purchase a replacement which does not suffer from the defect caused by the breach, or the cost of purchase would be greater than the cost of remedial work. Where damages are awarded by reference to the cost of remedial work, there is no requirement for the damages be spent on that work. But damages based on the cost of remedial work will not be awarded where there is a cheaper alternative which could make good the loss.

This chapter is directed towards the consideration of the limit on damages for dilapidations. It does not extend to a consideration of the limit on damages for breaches of any other covenants where *Tito* v *Waddel* could in many cases be in conflict with *Joyner* v *Weeks*.

Chapter 4

Compensation Where no Land is Taken

Preliminary

There are three ways of pursuing a claim for compensation where no land is taken, and none is easy:

1. Claim under the *McCarthy* Rules;
2. Claim under Part I of the Land Compensation Act 1973;
3. Start an action in Tort, from which the Authority may have statutory immunity.

Section 10 of the 1965 Act allows a claimant to refer matters of injurious affection, along with claims for the land taken, to the Lands Tribunal. Section 63 of the 1973 Act awards interest from the date of the claim until payment.

A fuller treatment of the material in this chapter together with worked valuation examples and summaries of the leading cases will be found in the author's *Handbook of Land Compensation*, published by Sweet and Maxwell.

1. The *McCarthy* Rules

These derive from *Metropolitan Board of Works* v *McCarthy* (1874) LR 7HL 243, which itself derived from *Chamberlain* v *West End of London & Crystal Palace Co* [1863] 2 B&S 617. In both cases the claimants suffered losses from the stopping up of public highways which were different from those suffered by the general public.

To claim, these four rules must be satisfied:

1. The loss must derive from the authority's action, made lawful by statute.
2. The loss would be actionable at common law were it not for the statute.
3. The claim must be for depreciation in the value of property and not for loss of trade. However, rent is a function of profit and the land's usefulness: if profits or utility decline, rent must

reduce. Capital value depends on rental value, and the rest
follows from this.
4.   The loss must arise from the physical execution of the works
      and not their subsequent use.

The fourth of these causes the most difficulty. It is the reciprocal of
Part I claims, where the loss must derive from the use and not the
construction.

Claimants who have had some land taken may claim their full
losses, including trade disturbance, measured by reference to
construction and use. It is, therefore, far better to have some land
taken than merely to sit close to a scheme of public works. The
surveyor who, by objecting at a compulsory purchase inquiry,
merely succeeds in driving the project across the claimant's
boundary fence, so that no land is taken, does that owner no service.

In *Wildtree Hotels* v *Harrow London Borough Council* [2000] 2 EGLR
5 a claim for loss of profitability attributable to a protracted road
scheme was attempted. The works reduced the hotel's accessibility,
and interfered with it by kicking up noise and dust. The Law Lords
agreed with Lord Hoffmann, who said:

> Damage to the amenity of the land caused by nuisances involving
> personal discomfort, having the effect of reducing the value of the
> land to let or sell, is damage to the land just as much as physical
> injury.

He then went on to say that such claims would normally founder
because it would be 'almost impossible' to satisfy the requirements
that the loss be caused by the construction of the works, made
lawful by statute, and actionable in tort but for this exemption.

As to liability in tort, there is a further fence to jump, because the
reasonable use of land is not actionable by neighbours. However, a
contractor who does not 'carry out work and conduct the
operations with all reasonable regard and care for the interests of
other people' will forfeit statutory immunity.

Damage caused by temporary interference has caused some
debate: brutally expressed, one argument is that since the horror
and the loss it causes both go away when the work is finished, what
is the complaint? Some valuers for compensating authorities have
put this forward – one hopes, tongue in cheek!

A valuer's response may well lie in the use of Term and
Reversion valuations, or dcf. It is not difficult to value the
diminished capital value as a perpetuity, and to add back the
difference between that and the full value once the works have been

completed, discounted for the period of the disturbance caused by the construction phase. For property which is let (or available for letting), a conventional term and reversion is sufficient: the reduced rent for the period of construction treated as the term[1], with reversion to the full F.R.V. once the works have been completed. Clearly, if the property is let, the terms of the actual lease must also be built into the calculation to reflect the contractual terms. A freeholder may, or may not, feel inclined to give the tenant some rental relief while the works are under construction; but if no relief is available for the tenant, the latter's claim for damage to the value of the leasehold interest could be calculated on the lines indicated above. Depending on the rent passing and the F.R.V., one may have to deal with such things as the valuation of negative profit rents. A lease held at the undisturbed F.R.V. moves into negativity if external works reduce the rental value below that level – from the lessee's point of view, rent paid will exceed rent received. It may be simpler to use a dcf approach under these circumstances as this removes the problem of applying a sinking fund to a negative income which the dual rate Y.P. raises. One does not need a sinking fund to replicate a loss!'

The Court of Appeal ruled, in *Clift* v *Welsh Office* [1998] RVR 305, upholding the Lands Tribunal, that compensation for physical damage to property caused by dirt and dust from construction sites is recoverable. In *Clift* the claim was for cleaning mud and dust from highway construction works from the inside and outside of a house. The court distinguished *Andreae* v *Selfridge & Co Ltd* [1938] Ch 1 on the basis that the latter referred to intangible loss and personal inconvenience, not physical damage.

In *Andreae* it had been held that everyone must put up with a certain amount of discomfort from neighbouring construction works, that the noise and dust were unavoidable, and were not actionable provided the operations were carried out reasonably and that steps were taken to avoid undue inconvenience.

---

[1] With no downward adjustment of the yield on account of this rent being less than the normal F.R.V., because this reduced rent is the F.R.V. while the work is in progress. One might argue that a higher yield should be adopted, to reflect the much greater difficulty in letting whilst the work is going on. Clearly, with a sitting tenant, there is no question of having to re-let; but vacant property adversely affected by the start of public works nearby is another matter.

Sir Christopher Slade, in *Clift* said:

> We see no sufficient reason why as a matter of policy the law should
> expect the neighbour, however patient, to put up with actual physical
> damage to his property in such circumstances. Where there is
> physical damage, the loss should fall on the doer of the works rather
> than his unfortunate neighbour.

Loss due to physical damage caused by the construction of works
– whether public or otherwise – is therefore recoverable regardless
of whether or not land is taken from the claimant. The author sees
no distinction in principle between construction work that causes
cracking in a nearby building and that which merely makes
everything filthy – and imposes cleaning costs in the process. In
Wildtree, the House of Lords concluded that there was no reason
why damage to amenity caused by nuisances involving physical
discomfort which reduced the value of the land was not damage to
the latter just as much as was physical injury.

2. Part I Claims: Land Compensation Act, 1973

Requirements

The availability of compensation under Part I is absolutely
dependent on the terms of the statute, and strict compliance is
essential for success. The facts of each case should be checked
against the wording of the Act. For instance, to claim successfully,
the claimant must have acquired an owner's interest before the
relevant date[2] and be in occupation at the date of notice of claim
(which can be between one and seven years after the relevant date)
– but dwellings and sales contracted between the relevant date and
the first claim day both provide exceptions. Although Part I is not
long, the detail is very finely drawn.

The claimant must satisfy these requirements:

- Have an interest of the right sort.
- Have the right sort of property.
- Have acquired it at the right time.
- Suffer from the right sort of depreciation.
- Caused by the right sort of nuisance.

---

[2]   The reader should remain undaunted: the terms used in this
paragraph are all explained later.

- Caused by the right sort of public works.
- Claim at the right time.

Critical dates

The Relevant Date: The date the works were first opened to the public or first used[3]. The claimant's qualifying interest must have been acquired before this date.

Mortgagees may claim, the compensation being treated as if it were the proceeds of sale. If property is vested in trustees, the trustees are deemed to be occupiers even though a beneficiary is in occupation, unless the beneficiary is a tenant for life. If property is jointly owned, all the joint tenants must claim[4].

The relevant date could be contentious if the scheme is developed in stages – see *Davies* v *Mid Glamorgan County Council* [1979] 38 P&CR 727, LT, where an airport was extended in three stages, each being completed and brought into use before the next. The Tribunal held that, when the project was looked at in its totality, the development amounted to a single scheme, requiring one claim, not three. Whether works which comes forward on the installment plan from one or more than one project for these purposes will depend upon the facts and is likely to be case-specific. In *Clarke* v *Shropshire County Council* [1996] (LT Ref/128/1994) public works were undertaken in two stages, under separate contracts funded by different derelict land grants. The preparatory work was in stage 1, the construction in stage 2. It was held that some of the works to which the claim related were within stage 1, and some within stage 2.

If the claim is in response to alterations to public works, the relevant date becomes:

---

[3]   Sections 1(9) and 9 of the 1973 Act.
[4]   Sections 10 and 11.

Carriageway alterations:
The date the highway is first opened after completion of the alteration.

Alterations other than to highways:
The date the works are first used after the completion of the alteration.

Changes of use of public works, excluding highways and airports:
The date of the change of use. An intensification of use does not count as an alteration[4].

The First Claim Day:	366 days after the Relevant Date[5]. It is the valuation date so far as prices are concerned. Claims must have been submitted within six years of this date or they will be out of time under the Limitation Act 1980. The Lands Tribunal is without jurisdiction to hear claims that are out of time[6].
Date of Notice of Claim:	This fixes the interest to be valued and the condition of the land to be valued[7]. Owners of qualifying leasehold interests should be advised to claim before the effluxion of time either diminishes the unexpired term's value, or, worse, erodes it below the minimum three-year threshold.

The claimant

Section 2

Save in the case of dwellings[8], the claimant must occupy the property at the date of notice of claim (not the relevant dated or first claim day). Occupation has the same meaning as in Rating law.

---

[4] Section 9(7).
[5] Section 3(2).
[6] Sections 3 and 19.
[7] Section 3(1).
[8] Regarding which please see later.

The claimant must be an owner who acquired the interest before the relevant date[9]. An exception is made for those who inherit property after the relevant date, provided the predecessor's interest qualified.

'Owner' for Part I claims means a freeholder or someone with a tenancy for a term of years certain (whether granted or extended) with at least three years unexpired at the date of the Notice of Claim.

It does not matter if the works were in progress at the time of purchase so long as this was before the Relevant Date, nor if a depreciated price was paid. The value is assessed on the First Claim Day, and if this gives compensation for depreciation to somebody who bought at a depreciated price, so be it: see *Fallows* v *Gateshead Metropolitan Borough Council* [1993] 66 P&CR 460, LT. This is statutory compensation: if the claimant has a qualifying interest and can prove depreciation due to the physical factors as on the First Claim Day, that is sufficient – the historic price paid for the interest is irrelevant.

A vendor who sells after the Relevant Date but before the First Claim Day at a depreciated price may secure Part I compensation by serving a Notice of Claim between contract and completion. This is a narrow window of opportunity, which, if taken, means that the compensation, when assessed on the First Claim Day, will be paid to the vendor. If not taken, no compensation will be recoverable because the purchaser from such a vendor will not be able to claim, regardless of whether or not the vendor had registered a claim, because the purchaser's interest would not have been bought before the relevant date, and so would not fit the definition of a qualifying interest.

Strict compliance with the rule is essential if the vendor's claim is not to fail. Some relief might be granted by the Tribunal if there has been some inadvertent mistake. In *Dodd* v *Stansted Airport Ltd* [1998] LT Ref/192/1996 the vendor registered a claim on 7 March in the belief that the contracts had been exchanged on the 3rd. In fact the purchaser did not sign his half of the contract until the 11th, although the vendor's solicitor had been led to think it had been signed on the 3rd. Some years later, when the compensation had been agreed at £25,000, the compensating authority's solicitor

---

[9]   It does not matter if the owner bought after the construction work started, nor if at a depreciated price. *Fallows* v *Gateshead Metropolitan Borough Council* (1993) 66 P&CR 460.

noticed the discrepancy, and sought to avoid payment because of it. By this stage matters had gone on beyond the point where things could have been corrected by the service of a second notice. The discrepancy in dates made no difference whatever to the valuations. The Tribunal held that the behaviour of the parties had created an estoppel by convention which prevented the council from doing anything of the sort: it had to pay up. Had the council wished to challenge the validity of the claim on such a technicality, it ought to have raised the challenge much sooner.

The same protection is available to those who have contracted to let non-residential property. Lettings of dwellings are excluded because the owner does not have to be in occupation – unless entitled to occupy – in order to make a Part I claim.

If the claimant is a tenant, it is important not to let the value slip away by delaying the claim. One has seven years from the Relevant Date[10], but delay reduces the unexpired term.

Dwellings and other property distinguished

Dwellings are treated differently.

Dwellings

Section 2(2) and (3).

The property does not have to be the claimant's only or main residence.

The claimant:

- Must have an owner's interest as described above, and
- Must occupy the dwelling if entitled to do so – which the freeholder of let property cannot.

Let property

Having let it does not bar the owner's claim. Only in the case of dwellings can two Part I claims subsist at the same time – one for the freeholder and one for the occupying tenant. The effect of a short letting agreement was at issue in *Allen* v *Department of Transport* [1994] 68 P&CR 347.

---

[10]   ie six from the First Claim Day – Statute of Limitation applies.

Vacant property

An owner who left the property unoccupied could not claim. The cure is to move back before serving the Notice of Claim. The same test of occupancy applies here as in Rating law.

Leasehold enfranchisement

The leaseholder will qualify if the Relevant Date is between the date of the Notice of Enfranchisement and the date the freehold or extended lease is acquired, regardless of the length of the unexpired term, provided the enfranchisement notice was served before the Relevant Date.

Any property except dwellings

Section 2(3)–(6).
   The claimant must be an owner occupier of

- An agricultural unit or
- Land in hereditament.

Land in hereditament

The annual value must not exceed £24,600[11]. If it does, no claim is allowed.
   An owner-occupier of land in hereditament is somebody who occupies the whole or a substantial part by virtue of an owner's interest.

Agricultural unit

An owner-occupier is somebody who is entitled to an owner's interest in at least part of it.

Short let property

Dwellings excepted, the owner must be in occupation at the date of

---

[11]   This is reviewed from time to time and is the same as in Blight Notice cases. This limit may be scrapped if the Law Commission's current [2002] proposals are accepted.

Notice of Claim. If possession cannot be regained within the limitation period of six years from the First Claim Day, there is a problem. If the tenant has three years unexpired, it is the tenant who is qualified to claim, but if the unexpired term is less, neither the tenant nor the landlord will be entitled to claim.

Depreciation

- Must be caused by the use of the works, not their physical presence. This makes the valuation difficult.
- It is far better to have a bit of land taken than not, as this allows a normal claim for injurious affection to be made. Such a claim admits of compensation for the full effect of the works, both as to construction and use.
- If the authority undertakes, or grant-aids, work to alleviate the effect of the works, the value of this benefit[12] is deducted from the compensation.
- Any betterment conferred by the works is deducted from the claim.
- Not all changes cause depreciation. It does not follow that things have got worse just because there has been a change to the status quo.

The physical factors

The depreciation must have been caused by one or more of these seven factors:

- Noise.
- Vibration.
- Artificial lighting.
- Smell.
- Smoke.
- Fumes.
- The discharge of solids or liquid onto the land in respect of which the claim is made.

Noise and fumes

See *D Jackson* v *Devon County Council* [1996] LT Ref/198/1993. The

---

[12]   NB the value, not the cost, of the alleviation work.

house in this case had frontage to a roundabout at the junction of four roads. The carriageways were altered, and widening resulted in part being slightly nearer the house. Traffic lights were installed. The property was put on the market at £285,000. The price was reduced several times, and it was eventually sold for £190,000. A considerable amount of evidence was produced as to traffic flow and noise levels. Overall, the authority's noise evidence was preferred to that of the claimant. The asking price sought, and the movement of the housing market, were complicating factors. The Tribunal accepted that the installation of traffic lights had caused a change in the behaviour of traffic at the roundabout, which could have increased exhaust emissions from standing vehicles. Lacking valuation evidence about the effect of increased fumes and dust, the Tribunal awarded £2,000.

Lighting on its own

See *Blower* v *Suffolk County Council* [1994] 2 EGLR 204. A listed building was affected by light from new street lamps. The amount of light arriving at the house was minimal, but the loom of the light in night sky was adverse when viewed from the terrace. Compensation was awarded. There is a distinction between noise and light: noise must arrive at the property with sufficient volume to affect value; light is seen from the property. The award did not breach the general rule that one does not have a right to a view – that was not the issue. What mattered was whether the claim satisfied the requirements of Part I – and it did: depreciation caused by artificial illumination emanating from the works is compensatable. A more common cause of complaint is, however, the loom of vehicle headlamps shining into bedrooms.

Noise and artificial lighting

Wakeley and Lambourne v *London Fire and Civil Defence Authority* [1996] LT Ref/45/1994. A fire station, for four appliances and accommodation for personnel, was brought into use in March 1988. The station responded to about 2,250 calls annually and was used for training operational crews, as required by law. The noise generated complaints. Only depreciation attributable to the noise and artificial lighting emanating from within the fire station premises was compensatable. Much evidence was offered about the levels of noise, disturbance and trends in value. As to noise, the compensating authority's expert argued that the increase in average hourly noise levels should be used (which increase was small), and that their irregular occurrence made them less detrimental. The claimants' expert said that the impact of specific fire station events (such as bells and sirens, changing the watches, training and other activities) should be considered as peaks set against the general level of background noise (which peaks were substantially different). On noise, the Tribunal held that the irregularity made things worse, that a prospective purchaser would not, perhaps, consider measured noise levels, nor be comforted (after hearing an exercise or call out) by being told that, mathematically, the average hourly noise levels were not much changed by them. The Tribunal agreed with the claimants, that purchasers would be more influenced by their own subjective judgment and thought it impossible there had been no diminution in value, and turned to the opinion evidence of the valuers. The claimants contended for a 33% reduction in value, the authority's valuer said there was no diminution. The Tribunal assessed this at the mid-point between 5% and 7.5%. On value, the Tribunal rejected calculations based on movements in building society house price indices taken over a wide area as being undependable. The comparable evidence offered was not found by the Tribunal to be of assistance, and opinion evidence was relied on. 60% was deducted from the vacant possession value of one house on account of a regulated tenancy.

The public works

Those works admitting Part I Claims are:

- New or altered highways.
- Aerodromes, unless occupied by a government department. Depreciation caused by aircraft leaving military airfields is not compensatable, but this is the only Crown Land exemption[13].
- Other works or land provided or used in the exercise of statutory power.

The physical factors causing the depreciation must have their origin in or on the public works. Noise from aircraft arriving or leaving an airport is an exception: it does not matter if the plane is within or beyond the airport boundary.

---

[13]   Section 84(1).

The reason for an increase in traffic on an altered road is irrelevant provided the factors have their source on the altered length and are caused by its use. The increase in traffic does not have to be a result of the alteration: see *Williamson* v *Cumbria County Council*, discussed below.

Alterations to public works

Section 9

What counts as an alteration is closely defined. The section deals with depreciation which would not have occurred but for the alteration. There are specific definitions of the relevant date applicable to claims following alterations: see under relevant date, above.

The following should be noted:

- Not all alterations are compensatable.
- The physical factors must emanate from the altered part.
- The Notice of Claim must specify the alteration or change of use responsible for the depreciation.
- An intensification of use by itself is not an alteration. There must be physical works of alteration.
- The betterment claw-back provision is restricted to betterment which is attributable to the alteration.

Carriageway alterations

The level, location or width must have been altered, and the depreciation must be caused by physical factors arising from the altered part. Motive is irrelevant: see *Williamson* v *Cumbria County Council* [1994] 2 EGLR 206 in which the authority lost the argument that the increase in traffic on an altered highway was due to a new road bridge further away and that the increase in traffic was unrelated to the alteration.

One of the most significant highway cases in recent years has been *King* v *Dorset County Council* [1997] 1 EGLR 245. In this case alterations to the carriageway were at issue, and in particular whether resurfacing amounted to an alteration – it did not. Substantial compensation was secured, largely because the claimants produced soundly argued and well-researched evidence to establish quantum.

Alterations to junctions, the installation of traffic lights, calming measures and mini-roundabouts have attracted attention. Whether there is depreciation is a matter of evidence and will vary from case to case, but it by no means follows that these alterations make things worse. A reduction in traffic speed has to be set against additional noise and fumes as vehicles slow, change down and accelerate away – and then converted into cash! *Underwood* v *Solihull Metropolitan Borough Council* [1997] LT Refs/214–217/96, 222/96 and 11/97 and *Lorimer* v *Solihull Metropolitan Borough Council* [1997] LT Refs/ 206–209/1996 were two such attempts. See also *D Jackson* v *Devon County Council* [1996] LT Ref/198/1993.

Statutory immunity

Sections 1(6) and 17

In highway cases, the highway must be one which is maintainable at the public expense[14]. The question of statutory immunity is irrelevant.

In all cases except highways, a Part I claim can only be made if the compensating authority is exempt from an action in Tort for nuisance – whether expressly or impliedly. The exemption must be conferred by an enactment which relates to the work in question.

In all non-highway cases, exemption from an action in nuisance is a crucial prerequisite. Without it, the claimant must proceed in Tort or go without compensation.

An authority wishing to avoid a Part I claim must establish that it has no statutory immunity in Tort. If it succeeds, it cannot subsequently plead immunity as a defence to an action in nuisance, even if it turns out it did have immunity after all.

Immunity was at issue in *Vickers* v *Dover District Council* [1993] 1 EGLR 193, LT. The council provided a car park[15] which adversely affected nearby flats. The council defeated a Part I claim by showing that the 1984 Act was permissive, not mandatory, and so gave no immunity from an action in nuisance. The judgments of the House of Lords in *Metropolitan Asylum District* v *Hill* (1874) LR 794 HL 193 and *Manchester Corporation* v *Farnworth* [1930] AC 171 were relied on. The same thing happened in *Marsh* v *Powys County Council* [1997] 2 EGLR 177 with reference to noise from a new primary school. The Education Act 1944 is permissive regarding the siting of schools, and confers no immunity.

The following judgments were cited in *Vickers*.

---

[14]   As defined in s 329(1) of the Highways Act 1980.
[15]   Under the Road Traffic Regulation Act 1984.

> Where the terms of the statute are not imperative, but permissive, when it is left to the discretion of the persons empowered to determine whether the general powers committed to them shall be put into execution or not, I think the fair inference is the Legislature intended that discretion to be exercised in strict conformity with private rights, and did not intend to confer licence to permit nuisance in any place which might be selected for the purpose... These powers appear to me to be from first to last permissive and not imperative. Whether they shall be exercised at all, and, if so, to what extent and effect their exercise shall be carried out, is left to the discretion of the Local Government Board.
>
> [Lord Watson in *Metropolitan Asylum District*]

> When Parliament has authorised a certain thing to be made or done in a certain place there can be no action for nuisance caused by the making or doing of that thing if the nuisance is the inevitable result of the making or doing so authorised. The onus of proving that the result is inevitable is on those who wish to escape liability for nuisance, but the criterion of inevitability is not what is theoretically possible but what is possible according to the state of scientific knowledge at the time, having also in view a certain common sense appreciation, which cannot be rigidly defined, of practical feasibility in view of a situation and of expense.
>
> [Viscount Dunedin in *Manchester Corporation*]

If an Act is permissive, it does not carry an implied exemption from action, and statutory powers must be exercised in strict conformity with private rights. A body is liable for nuisance due to the exercise of its powers if it is either expressly made liable or is not exempted from liability.

All is not gloom. If an action in nuisance succeeds, the claimant can seek damages having regard to construction and use, and to anything else which is material, and may even ask for an injunction. Potentially, this promises fuller compensation than under Part I where the claim is limited to depreciation caused by the physical factors.

Making the claim

It is possible to lose the case by getting this wrong.

- Section 3 specifies what must be included in the particulars of claim. Failure to quantify the amount will sink the ship – see *Fennessy* v *London City Airport* [1995] 2 EGLR 167 in which the claim was stated as being for an amount in excess of £50. This was held to be insufficient to rank as a 'particular of claim'.

- The amount must exceed £50[16].
- 366 days must elapse after the Relevant Date. The only exception is that a claim may be made between the Relevant Date and the First Claim Day by an owner who has contracted to sell a dwelling (or to sell or let non-residential property) provided the transaction has not been completed at the date of claim.
- The claim must be submitted within six years of the First Claim Day.
- If the property is jointly owned, all joint tenants must claim.
- Only one claim may subsist at a time – once compensation becomes payable, no other claims may be made whether in respect of that interest or any other. Dwellings are the exception, where two claims may subsist (the freeholder's and the occupier's, provided their interests are qualifying ones)[17].
- If a claim fails to produce compensation, for whatever reason, the owner of any other qualifying interest may claim.
- A successful claim by an occupying tenant can (dwellings excepted) bar a claim for depreciation to the value of the free-hold reversion – a point to watch if there is a chance of getting possession prior to claiming as an occupying freeholder.
- Specified alterations to public works will unlock the possibility of further claims, but only in connection with the alteration.
- The duplication of compensation for injurious affection under section 7 of the 1965 Act is barred.
- If an Authority buys part of a vendor's holding for public works, no Part I compensation is payable on the land retained, in relation to those works, regardless of whether injurious affection is claimed or not.

The compensating authority

This, the responsible authority, will be the appropriate highway authority. In non-highway cases, it is the person managing the works[18].

---

[16] This is likely to be raised if the Law Commission's current recommendations are accepted.

[17] Section 9.

[18] Sections 1(4) and 19(1).

Interest

Interest at the statutory rates[19] is payable on the compensation from the later of the First Claim Day and the date on which the notice of claim is served.

Expenses of those moving temporarily

Section 28 allows, but does not compel, an authority to pay the reasonable expenses incurred by an occupier's household if it flees to alternative accommodation. The dwelling must adjoin the works, and its enjoyment must be affected to the extent that continuing occupation is not reasonably practicable.

The valuation rules

These are as follows:

- The First Claim Day is the valuation date.
- The interest and the condition of the property are as on the date of claim.
- Rules 2–4 of section 5, 1961 Act, apply[20].
- The effect of any mortgage and of contracts for sale or letting made after the Relevant Date are ignored[21].
- Schedule 3 consent, subject to Schedule 10, is to be assumed.
- Planning permissions which have not been implemented are to be ignored[22].
- Depreciation due to the physical factors is to be assessed having regard to the anticipated subsequent intensification of use as forecast on the First Claim Day. *Bwllfa and Merthyr Dare Steam Colleries (1891)* v *Pontypridd Waterworks Co* [1903] AC 426 cannot be directly applied to allow the substitution of accurate data because the statute defines both the valuation date and the assumptions[23]. The principle can be used to admit comment on the accuracy and reliability of those forecasts, and of the weight

---

[19] See the current Acquisition of Land (Rates of Interest after Entry) Regulations.

[20] Section 4(4).

[21] Section 4(4).

[22] Section 5.

[23] Sections 3(2)(3), 4(1). See *Dhenin* v *Department of Transport* (1989) 60 P&CR 349.

to be attached to them.

- Only the effect of physical factors emanating from within the boundaries of the works is to be assessed, but aircraft noise is treated differently.
- Set-off is to be deducted for betterment given by the construction or use of the works, whether benefiting the reference land or other contiguous or adjacent land held by the claimant in the same capacity as the reference land.
- The value of soundproofing work by the Authority (or grant aided), and of any mitigating works carried out on other land is to be deducted from the depreciation. It is the added value, not the cost of the work, which is to be deducted.
- Some remedial works are mandatory. It is to be assumed any mandatory soundproofing works have been carried out, or the grant paid, whether or not this has happened. If the works are discretionary, they are only to be assumed if the authority has undertaken to exercise its discretion[24].
- Interest is recoverable from the later of the First Claim Day and the date of claim.
- Professional fees are recoverable. The Tribunal has said it should not normally be necessary to call two experts – one local valuer and another to explain how to split the depreciation between physical presence and use is one valuer too many.

Valuations

Each case is different. Here are some pointers:

- Hard market evidence based on comparables is best.
- One can work from either the Depreciated or Undepreciated figures, depending on what is available; one may use a Before and After approach, or a With-Construction-and-Use less a With-Construction-but-No-Use analysis. All have been tried. The author favours working from a No-Scheme approach, assessing the total injurious affection and then apportioning between construction and use. What is clear is, thankfully, that there is no one 'correct' method.
- The Lands Tribunal regards evidence based on other settlements with a 'mildly benevolent eye'[25] – they are not

---

[24]   Section 4(3).
[25]   The Member in *Marchant* v *Secretary of State for Transport* (1979) 250 EG 559.

market evidence, only sums which claimants were prepared to accept. Settlement evidence may provide a cross-check on opinion evidence and/or the accuracy of analysis. Were the claimants professionally represented, or did they simply accept what was on offer? The Tribunal in *Arkell* v *Department of Transport* [1983] 2 EGLR 181 found settlement evidence of little help, a view reiterated in *Nesbitt* v *National Assembly for Wales* [2002] (LT LCA/139/2001). In the latter, the Member drew attention to the remarks of RC Walmsley, FRICS, the Member in *Farr* v *Millersons Investments Ltd* (1971) 22 P&CR 1061, who said:

> Unless the settlement evidence is shown to provide solid support, the Tribunal attaches little weight to it: for instance if the tenants who settled had done so without professional advice, or despite such advice; or if there is clear evidence as to the basis on which the settlements were negotiated; or if the valuer producing the settlement evidence was not personally concerned in the negotiations; or if there is market evidence which puts in doubt the site value contended for; then in any or all of these circumstances the evidence afforded by settlements is readily displaced by other evidence.

- It is always wrong to lift percentage allowances from other cases. The percentages are case-specific, and the fact that Property A in Town B was depreciated by n% by such-and-such works at m metres, generating d decibels on a wet Thursday does not indicate that n% is the correct figure for Property D in Town F. One is expected to use one's own professional judgment, and to back opinion with market evidence if any can be found. See the Member's remarks in *Fallows* v *Gateshead Metropolitan Borough Council* (1993) 66 P&CR 460:

> ... I would comment, however, that the facts in this type of case are likely to vary so much in respect of such matters as the size and type of highway, distance from the property concerned, traffic intensity, etc., that it is highly unlikely that any pattern can emerge which can be relied on for evidential purposes.

- The indexation of values is frowned upon. The index indicates market trends in general terms, and can act as a check, but one cannot value an individual property directly using indices. Market changes over a specific period should be tracked by reference to local comparables.
- Scientific evidence has been welcomed, and failure to know the noise levels etc. weakens one's case, but science is not the

determinant of value. Opinion, experience and rational judgment need to be applied when considering and weighing up what the scientific instruments have to say. The effect the physical factors have on purchasers depends on their perceptions, not the position of a needle on a noise machine.

- The cases show that the Lands Tribunal is unconvinced that there is really a strict or mathematical correlation between decibels and depreciation – prospective purchasers use their ears and eyes, not little black boxes.

- The better the property; the fussier the purchaser. Thus, the greater the allowance needed to lure the fish into taking the hook.

- The burden of proof is on the claimant, who must establish the loss: see *Shepherd and Shepherd* v *Lancashire County Council* (1977) 33 P&CR 296 and *Casewell* v *Department of Transport* (1984) 269 EG 1051 – A Leasehold Reform Act case, but apposite.

- The burden of establishing any set-off is on the authority: see *Durnford* v *Avon County Council* [1994] LT Ref/45, 46, 142–146/ 1993.

- When valuing for a claim following the alteration of public works, it is the increase in the physical factors emanating from the altered part of the works which have to be appraised. This calls for the wisdom of Solomon and a steady nerve as one slices the depreciation due to this away from everything else. If the works pre-date the 1973 Act, the initial depreciation will remain outwith the compensation provisions, if they post-date it, the initial depreciation should already have been claimed in its own right by whoever was entitled to do so at the time.

- A significant number of claims fail because of incompetent valuations. Some errors are crass, for example using asking prices and estate agents' particulars as evidence of market value.

- Other claims fail because the claimant cannot prove there has been any financial depreciation. There is a difference between different and worse when looking at the effect of public works. See *Hickmott* v *Dorset County Council* [1977] 2 EGLR 15, CA; [1975] 1 EGLR 166, LT.

- The claimant's own perceptions are accepted, but treated with caution unless there is some independent, objective, evidence too.

- Compensating authorities seem apt to make up their own rules – which one can dispute lustily. For example, 'we do not pay if the property is more than xyz yards from the road': *Marchant* v *Secretary of State for Transport* [1979] 1 EGLR 194, or 'We measure noise by reference to our own datum and do not pay unless it is exceeded': *Jenks* v *Northamptonshire County Council* (1993) 66 P&CR 303. The authorities lost in both instances.

* Just because the depreciation is high, or a lot higher than the authority is used to, does not mean it is wrong. But prove it! See *King* v *Dorset County Council* [1997] 1 EGLR 245.

Noise payments for occupiers of moveable homes

Section 20A and Regulations

Qualifying occupiers are entitled to claim for a noise payment of £1,650 from responsible authorities (as defined earlier in this chapter), provided prescribed conditions are satisfied. Different Regulations apply in England and Wales, and some of the differences in wording are material[26].

A moveable home is defined as:

* A boat designed or adapted for use as a permanent home, lawfully moored with whatever consent is needed from the navigation authority and with the landowner's written consent.
* A caravan, excluding a motor caravan, as defined by the Caravan Sites and Control of Development Act 1960 and section 13 (disregarding 13(2)) of the Caravan Sites Act 1968, lawfully stationed on a protected site throughout the qualifying period.

Two qualifying periods are defined: 18 months for noise from construction and three years for traffic noise after construction. The claimant must have occupied the moveable home as the only or main residence throughout the qualifying period. Regulations specify key dates, and these are different as between England and Wales.

A qualifying claimant may opt for compensation for either the construction of the works, or for the noise once the works have opened, but not for both.

So far as a claim for noise from the use is concerned, the regulations state that the noise must exceed levels specified in the regulations.

---

[26] Currently, see SI 2000 No 3086 for England, and SI 2001 No 604 for Wales.

So far as noise from the use of the construction is concerned, the test is whether the authority thinks the effect of the noise to be seriously adverse so far as the enjoyment of the home is concerned.

The difference between the English and Welsh regulations so far as noise-from-construction claims are concerned is significant.

- In England: The noise has had or will have a seriously adverse effect on the enjoyment of the home. The home has to be adjacent to the works.
- In Wales: The noise has had a serious effect, etc. There is no 'will have' in the Welsh Regulations, but the home does not have to be adjacent – it must be within 300 metres of the nearest point of the carriageway. NB Carriageway, not highway.

Claims must be submitted in writing within six years of the ending of the qualifying period, and must supply ten specific pieces of information, plus a declaration that what is stated is correct. There is no set pro-forma for the claim – a letter will do.

The highway authority may make a noise payment if the road becomes maintainable at the public expense within three years of the relevant date.

Chapter 5

Rights to Light

Introduction

The issue of rights to light is an area that is frequently overlooked by parties both with an existing relevant interest in land and those seeking to acquire a relevant interest in land. Rights to light, in England, Wales and Northern Ireland (they do not apply in Scotland), have always been an important consideration in the development process particularly so within the urban environment. The move towards restricting greenfield and promoting brownfield development in recent years, a trend which is likely to continue, has reinforced both the significance and importance of rights to light.

Developers acquiring a site without considering potential rights to light issues may well have to make potentially large unexpected compensation payments to adjacent owners and in some cases may also have to reduce the scale of the proposed development with the consequential impact on profitability and viability, not only of the particular scheme but also, potentially, of the development company.

However, rights to light are not and should not be regarded as the exclusive concern of developers. Property owners and occupiers, and their consultants, need to be aware of rights to light and that the redevelopment of adjacent properties may trigger a claim for compensation or enable other gains and/or rights to be achieved by negotiation. Furthermore, owners whose properties have long-term redevelopment potential may, in some cases, be able to prevent neighbouring properties from acquiring rights of light over their properties, thereby protecting and potentially enhancing the long-term development potential and value of their property. This chapter aims to give the reader a basic introduction to rights to light; identifying what they are; how they can arise; how they can be prevented and/or extinguished; considering the potential implications when rights of light are infringed and giving a brief explanation as to how rights to light are assessed in terms of value.

What is a right to light?

A right to light is an easement, namely a right which one area of land (the dominant tenement) has over another area of land (the servient tenement). It is not appropriate within the context of this chapter to provide a full explanation as to the nature of easements. There are numerous authoritative texts on easements, in particular *Gale on Easements*, 17th ed, edited by Jonathan Gaunt QC and Paul Morgan QC (Sweet & Maxwell). For a fuller discussion as to the nature of the easement of light the reader is referred to *Rights of Light – The Modern Law* by Stephen Bickford-Smith, Andrew Francis, and Elizabeth de Burgh Sidley (Jordans).

In considering rights to light the basic points to bear in mind are as follows:

(i)   A right to light is a negative easement in that it is a right whereby an owner of land may be able to prevent the owner of other land from utilising that other land in a particular way.
(ii) It is a right that attaches to a particular window aperture [or light] within a building.
(iii) The grant of planning permission does not extinguish any rights of light that may exist.

How rights to light arise

In broad terms rights to light can be acquired in one of two principal ways: the first is by contract and the second is by prescription under the Prescription Act 1832. (There are other ways, as for example under the doctrine of lost modern grant, but these do not concern us other than to note that the method of valuation is usually the same.)

Contract

A contractual right to light can be acquired by one of two main avenues, express grant or by implication and can, depending on the precise wording of the relevant deed, be either restrictive or permissive in nature. If the contractual right to light is permissive then, typically, the parties to the deed will be permitted to develop their land up to an agreed parameter or to an agreed benchmark, for example in accordance with any approved planning permission, but are not contractually bound to keep within such limits. If the contractual right to light is restrictive then any permitted development may only be in accordance with specified plans or to a specific building envelope and no further or not otherwise. Clearly where a proposed development is in breach of a restrictive contractual right to light then the potential negotiating position of the injured party may be very strong. It should be noted that the examples shown above give a clear distinction between a permissive and restrictive contractual right to light. In practice the distinction is often unclear and the precise interpretation of the relevant deed or agreement may need to be referred to counsel.

A key issue for practitioners to bear in mind when considering potential rights of light issues is the importance of studying title documentation. It is frequently the case that property owners are not aware of contractual rights of light that exist even though such rights are specifically referred to on the office copy for their particular title. Furthermore, it is important to consider office copies and title documentation in respect of adjoining sites in addition to the particular site under consideration. There are invariably at least two parties to any contractual rights to light documentation and instances do occur where the documentation is only referred to on one of the relevant office copies. An examination of office copies for adjoining sites may often identify contractual rights of light which were previously unknown.

Rights to light can also be acquired by implication under, for example, the terms of a conveyance and section 62 Law of Property Act 1925.

Prescription

Prescriptive rights to light arise under the Prescription Act 1832 and are acquired after the relevant window apertures have been in place for a continuous period of 20 years. If an obstruction is caused for a period of not less than one full year, the prescriptive period is interrupted and starts afresh when the obstruction is removed. The corollary of this is that, once the window aperture has been in place and has enjoyed the benefit of a right to light for a continuous period of 19 years and a day, the acquisition of a prescriptive right cannot be defeated.

Contractual rights will generally take precedence over prescriptive rights to light. Hence window apertures that have not been in place for sufficient time to have acquired prescriptive rights may enjoy the benefit of contractual rights to light.

It is possible for a new building, with new windows, to have carried forward prescriptive rights to light from a building previously on the site. If window apertures in a new building or an aperture which has not yet been in existence sufficient time to have acquired a prescriptive right of light can be shown to be substantially coincidental with window apertures in a previous building on the site, then the new window apertures may well have the benefit of any prescriptive rights to light which were held by the previous window apertures.

In summary in considering whether or not window apertures on a particular building will have rights to light over a particular site it is not sufficient merely to rely on the age of the relevant building. Windows which have been in place for less than 20 years can have the benefit of prescriptive rights of light and even if it can be shown that they do not have prescriptive rights of light they may have the benefit of contractual rights to light. The importance of researching the history of and documentation relating to the relevant properties cannot be over emphasised.

Extinguishment of rights to light and preventing acquisition

Extinguishment

A right to light can be extinguished in one, or more, of the following ways:

(i)   By agreement.
(ii)   By unity of ownership.
(iii) By permanent abandonment.
(iv)   By interruption.
(v)   By demolition or alteration of the dominant building.
(vi)   Under statutory provisions.

By agreement

A right to light, whether contractual or prescriptive in origin, may be extinguished by agreement between the relevant parties. It is important that any such agreement is properly documented and that the document covers and includes all those parties with a relevant interest. For example an agreement executed solely by Freeholder A and Freeholder B which extinguishes all rights of light between their respective properties may not necessarily extinguish rights to light enjoyed by their respective tenants over the other's property.

By unity of ownership

The right to light is an easement and it is an established principle that an owner cannot have an easement over his own property. Any rights of light enjoyed by Freehold A over Freehold B would be extinguished should the respective titles be unified. It should be noted however that tenants may acquire rights of light over the property of their landlords which may not necessarily be automatically extinguished by this method.

By permanent abandonment

If it can be established that the owners of the dominant tenement have, by their action, shown a clear intention to permanently give up any right to light to which they may be entitled, then the right may be extinguished. There must however be a clear intention to permanently give up the right. For example merely bricking up a window opening, even for a considerable period of time, may not necessarily signify abandonment of any right to light enjoyed by that window aperture. From the above it can be seen that it is often very difficult to establish 'permanent abandonment' in practice.

By interruption

Under the Prescription Act 1832 an interruption to a right of light for more than one year effectively extinguishes that right (although it does not prevent it from being acquired again). The interruption can be either physical or notional and the dominant owner must have acquiesced in the interruption. A physical obstruction will normally take the form of a new structure, whilst a notional obstruction can only be in the form of a Light Obstruction Notice served in accordance with the provisions of the Rights of Light Act 1959. Such notices are served by the servient owner on the dominant owner and specify a notional structure on the servient land of either a specified height and envelope or an unspecified height. Light Obstruction Notices once registered remain in force for one year after which they cease to be effective. Consequently, if the intention is to permanently extinguish rights of light, Light Obstruction Notices will need to be served on a regular basis to ensure that the right is not enjoyed for any subsequent continuous period of 19 years and one day. There are specific procedures for the service and registration of light obstruction notices and it is

essential that these are properly adhered to as if not the Light
Obstruction Notice may be ineffective and will neither extinguish
nor prevent the acquisition of rights to light by the dominant
tenement. If there are good grounds to resist a Light Obstruction
Notice, (eg it can be proved that window apertures have enjoyed
light for at least 19 years and a day), legal action to challenge the
Light Obstruction Notice must be taken before the notice has been
on the register for a full year.

By demolition or alteration of the dominant building

If the building in which the apertures are situated which have the
benefit of any rights to light is demolished, then clearly a question
must arise as to whether or not the right to light continues to exist.
The situation with demolition is similar to that of abandonment
and the demolition of a building does not automatically result in
the extinguishment of any rights of light which the apertures in that
building may have had over any adjoining property. If it can be
shown that there is an intention to redevelop with a building which
has apertures which are substantially coincidental with the
apertures in the original building, then any rights of light enjoyed
by the apertures in the original building are likely to be transferred
and remain in existence for the benefit of the apertures in the new
building. It follows from this that it cannot necessarily be assumed
that cleared sites have no rights of light. Once again the importance
of investigating the history and title documentation of the relevant
sites cannot be over emphasised when considering rights to light
issues.

Under statutory provisions

Under section 237 of the Town and Country Planning Act 1990 local
authorities have the right to appropriate rights of light for planning
purposes. It should be noted that this right not only applies to
development by the local planning authority but also to any person
deriving title from it. Powers for the extinguishment of rights of
light are also contained in compulsory purchase legislation.

Prevention

The acquisition of rights to light can be prevented in one or more of
the following ways:

(i)   By agreement.
(ii) By interruption.

By agreement

The respective owners of the dominant and servient tenements may agree that neither will acquire rights of light against the other. Such an agreement can be drafted in a specific deed or may be included in another document. For example a part conveyance may often reserve full redevelopment rights for the vendor in respect of his retained land and for the purchaser of the property sold, or it may reserve such rights solely for one of the relevant parties. Also a lease may prevent the acquisition of any rights of light by a tenant over any adjoining or neighbouring premises of the landlord or its successors in title. As indicated in the section 'How rights to light arise' (p 78), an agreement may not always totally prevent the acquisition of rights to light and may permit either on a reciprocal or single party basis a development in accordance with a specified benchmark, for example specific plans or a general building envelope. It is vitally important that such agreements are well drafted and clearly and unequivocally state the intention of the parties. If agreements refer to specific plans then copies of those plans should be attached to the documentation. It is often the case that such agreements are poorly drafted and require counsel's opinion in order to arrive at a precise interpretation.

By interruption

As indicated in the section on 'Extinguishment' (p 80) above the acquisition of a prescriptive right to light may be prevented by an interruption of more than one year. The interruption may be physical, typically a new structure, or in the form of a Light Obstruction Notice served under the Rights to Light Act 1959. For further basic information on Light Obstruction Notices the reader should refer back to pp 81–82 above. More detailed information can be found in the books referred to in 'What is a right to light?' (p 78).

Extinguishment and prevention – general comments

Contractual rights of light can, in the main, only be extinguished or modified by agreement, unity of ownership or statute. Interruption, either physical or notional, will not usually have any impact on a contractual right of light. However it may be possible to argue that a dominant owner has given up any contractual right to light if he or she permits an obstruction to remain in place unchallenged for a suitable period. Prescriptive rights of light can be extinguished, modified or prevented by all of the methods set out above although abandonment can be difficult to establish in practice. Undoubtedly the key issue for practitioners to bear in mind is that rights to light need not necessarily be permanent. They can often be extinguished or prevented without the agreement of the owner of the dominant tenement and without the payment of any compensation. A pro-active management approach is required to ensure that potential development sites are not blighted by allowing the acquisition of prescriptive rights to light by neighbouring properties.

Infringements – the potential implications

The potential implications when rights to light are infringed are, to a degree, dependent upon whether the particular right to light is contractual or prescriptive in origin and also the extent of any infringement. If the right to light is contractual in origin the potential implications of an infringement will, in the main, be determined by the precise wording of the relevant documentation and the key issue in this regard is whether or not the right is restrictive or permissive. Where there is an infringement of a restrictive contractual right to light there is a strong possibility that the injured party may be able to obtain an injunction preventing the infringement. In such circumstances the negotiating position of the injured party will be very strong and any payment which they accept in lieu of an injunction is likely to be substantial. Alternatively, if the injured party is not prepared to accept a financial settlement, the design of any development may need to be amended to prevent the infringement. If site works have commenced the cost of redesigning a development, particularly a major redesign, can prove very substantial in terms of cost, time delays, and a potential profit reduction.

In the case of permissive contractual or prescriptive rights to light the potential implications of an infringement are either damages and/or an injunction. The remedy awarded by the Court in the first instance for any infringement should be an injunction but damages in lieu of an injunction may be awarded if the four tests laid down in *Shelfer* v *City London Electric Lighting Co Ltd* (1895) 1 Ch 287 CA are met. These tests are summarised below:

(a) Will the injury to the claimants' rights be small?
(b) Is the issue one which is capable of being estimated in money?
(c) Can the injury be adequately compensated by a small money payment?
(d) Would it be oppressive to the defendant to grant an injunction?

It should be noted that the difference between an injunctable and a non-injunctable interference is indistinct and in negotiation the potential risks of an injunction being secured may have to be assessed and reflected in any subsequent settlement. Even if the risks of an injured party securing an injunction are considered to be comparatively small, the potential delays and costs of taking a case to court will often justify a payment of an 'inflated settlement'. Where it is agreed that damages are appropriate the level will be dependent upon the extent of the infringement. However, six and even seven figure settlements are not uncommon especially where there is a significant infringement and a strong likelihood that the injured party would be able to obtain an injunction should they wish to do so. In such circumstances settlements can be based on a notional share of the development profit.

In cases where an injunction is awarded the injunction may take the form of either a prohibitory injunction, which would prevent the development from proceeding, or a mandatory injunction, which may require a completed development or structure to be removed or altered either in part or in its entirety.

In summary, the implications of infringing a right to light are potentially very significant and can, in certain cases, lead to the payment of very substantial levels of compensation and/or the total or partial redesign of a completed development. In short the consequences of ignoring potential rights of light issues can both from a developer's and an adjacent owner's point of view prove to be a very serious error of judgment.

How rights of light are measured and valued

Rights to light are measured in terms of light received directly from the sky over a neighbouring property. Whether or not a room or an area of a particular property is well lit, from a rights to light viewpoint, does not depend on the overall quantum of light that the room may receive direct from the sky and indirectly by reflection, but rather on the amount of sky which is directly visible at certain points in that room. This is because the quantum of light in any particular room is dependent upon, *inter alia*, the time of year, the time of day, the climatic conditions and the amount, if any, of artificial light available. The amount of sky which is visible at any point in a particular room is dependent upon the size and shape of the window aperture or apertures which serve that particular room, the configuration of that room and the location and size of any structures visible from the particular point through the relevant apertures. Rights of light practitioners refer to sky visibility. Levels of sky visibility are dependent solely on the design/layout of the relevant accommodation and, specifically, the scale and massing of nearby relevant structures; they are not affected by variables such as the time of day, the time of year or the weather.

Given that the concept of 'sky visibility' is now, hopefully, understood it is necessary to consider how much of the sky needs to be visible in order for a particular point in the room to be considered well lit. Scientific investigation has established that from a uniform overcast sky the amount of light from 0.2% or 1/500th of the total dome of the sky is sufficient for normal purposes. This benchmark of 0.2% sky visibility is generally accepted by rights to light practitioners. The level of sky visibility from any particular point in the room will be dependent upon the relative height of that point to the levels of the cill and the head of the window aperture or apertures through which the sky is visible and the massing of any obstruction visible from the relevant point through the window. To ensure a uniform approach measurements of sky visibility are taken at a constant height known as the working plane which is at 850 millimetres above the floor level of the particular room. The working plane is intended to correspond generally to desk top/table top/work bench level.

The process of identifying those points in a particular room that benefit from a minimum level of 0.2% sky visibility involves a complex analytical process based on what is known as a Waldram diagram.

Once those points with a minimum level of 0.2% sky visibility have been identified they are plotted on a plan of the relevant room; a 0.2% sky visibility contour is drawn and the area within the contour calculated. This process is repeated on a pre and post development basis in respect of each of the relevant areas or rooms affected by the development. The differences in the areas defined by the 0.2% sky visibility contour pre and post development show the loss, or gain, of 0.2% sky visibility arising from the proposed development.

The raw data obtained from this analysis, namely the areas of 0.2% sky visibility loss or gain, need to be presented in such a way to enable meaningful conclusions to be made. The process adopted by rights to light practitioners is to compare the areas lit to 0.2% sky visibility or more with the total area of the particular room. The generally accepted threshold that defines whether or not a room used for commercial purposes is well lit is 50%. Hence if 50% of the total area of the room benefits from 0.2% sky visibility or more then that room is regarded as well lit. Consequently if, for example, a development reduces the level of 0.2% sky visibility in a particular room from, say, 65% to 53% of the total room area then the particular room would, post development, be still well lit and would not have suffered any compensatable injury. It follows from the above that parties who suffer loss of sky visibility as a result of a particular development may not have an automatic entitlement to all the sky visibility which they currently receive over a particular property. This is an important principle and one that has been established in the cases of *Colls* v *Home & Colonial Stores Ltd* [1904] AC 179 and *Allen* v *Greenwood* [1979] 1 All ER 819, CA. It should be noted that in cases involving losses in residential property, the courts have indicated that the standard for a well-lit room should be greater than 50%, but no precise percentage figure has been laid down, each case being judged on its merits.

Whether or not a loss of sky visibility is potentially serious is established by zoning the area of loss. The loss is zoned by dividing the relevant room area into four zones of equal area which are referred to by rights to light practitioners as the front zone, first zone, second zone and makeweight zone. The front zone corresponds to the first 25% of the total room area; the first zone corresponds to the next 25% (namely 26%–50%) of the total room area; the second zone corresponds to the next 25% (namely 51%–75%) of the total room area while the makeweight zone corresponds to the final 25% (namely 76%–100%) of the total room area. Losses of sky visibility to either the second or makeweight zones would not, in isolation, be considered compensatable if the whole of the front and first zones, which in total amount to 50% of the total room area, remain well lit. However if there is loss to the front or first zone then less than 50% of the total room will be well lit and the 0.2% sky visibility threshold of 50% of the total room area will have been breached. Once front or first zone loss is established then losses to the second zone and makeweight zones become relevant.

Clearly a loss of sky visibility to the first zone of a room that still retains 0.2% sky visibility to over 25% of its total area, or the whole of the front zone, is not as serious as a loss of sky visibility to the front zone of a room. Losses of sky visibility to the front zone are considered to be the most serious and are therefore enhanced by a factor of 1.5. Losses to the first zone are taken to be not as serious as front zone loss but still serious and are taken at par. Losses to the second zone and makeweight zones are not significant in isolation but are taken into account once front and/or first zone loss has been established. Losses to the second zone and makeweight zones are reduced, respectively, by factors of 0.5 and 0.75. The adjustments that are made to the areas of loss in the front, second and makeweight zones enable the total area of loss to be stated on an 'Equivalent First Zone basis' (EFZ). The area of loss expressed in terms of EFZ is used by rights to light practitioners as a basis for calculating any appropriate compensation payments.

Unless there is a very serious loss of sky visibility the generally accepted approach is to assess the value or base the compensation on the area of loss as expressed in EFZ terms.

Once the area of loss in EFZ terms has been established, a percentage of the open market rental value is applied to it. It is generally accepted at the present time by rights to light practitioners that no more than £53.82 per m² (£5 per sq ft) of the annual rental value represents the value of light. Furthermore the percentage of the rental value attributable to light generally decreases with an increase in the rental value. As a general guide only, in a property with a rental value of up to £32.29 per m² (£3 per sq ft) around 50% or £16.15 per m² (£1.50 per sq ft) may be attributable to light. This percentage would decrease on a uniform basis to a minimum figure of between 25%–30% for rental values of around £75.35 per m² (£7 per sq ft) and above. Applying these general figures would therefore result in a light rental value of between £40.36 per m² and £48.44 per m² (£3.75 and £4.50 per sq ft) for a property with a rental value of £161.46 per m² (£15 per sq ft) and properties with a rental value of £215.29 per m² (£20 per sq ft) and above would have a light rental value of £53.82 per m² (£5 per sq ft), being the previously stated and generally accepted current maximum light rental value. Once established the appropriate light rental value is multiplied by the loss in EFZ terms to arrive at a total light annual rental value. This figure is then capitalised by an appropriate yield or years' purchase factor to give the base compensation figure or 'book value'.

In cases where there is little or no front zone loss the figure calculated by capitalising the light annual rental value would be the appropriate compensation figure. However, in cases where there is front zone loss it is generally accepted that the injured party may have a stronger negotiating position and be able to command a potentially increased compensation figure. This approach was established in *Carr-Saunders* v *Dick McNeil Associates Ltd* [1986] 2 EGLR 181. In that case levels of 0.2% sky visibility in two rooms were reduced from 55% to 4.5% and from 57% to 6.5%. There was clearly a significant element of front and first zone loss and it was agreed that the appropriate level of damages was around £3,000 when assessed on the conventional approach. However to take account of the potentially strong negotiation position of the injured party the court awarded damages of £8,000 which represented an uplift factor of around 2.67. A similar approach is now often adopted by rights to light practitioners in cases where front zone loss occurs.

In cases where there is a very serious loss and where the injured party is left with very limited levels of sky visibility taking into account the nature of the relevant property and its locality, the level of compensation is sometimes assessed as a percentage of the development profit that will accrue to the developer from that part of the development that will cause the injury. Such an approach may only be appropriate, and then by no means invariably, when the injury is serious enough to merit the granting of an injunction. Settlements in such cases often reflect an element of horse dealing.

Valuation of a right to light – a practical approach

Given that the preceding sections have dealt with the theoretical aspects of right to light, it is now appropriate to consider how rights to light issues are generally dealt with in practice. The approach set out below considers the 'developer's' viewpoint, but the approach is equally appropriate when acting for a potentially injured party although clearly there will be some differences in emphasis. The valuation of rights to light is, essentially, a sequential exercise that can be divided into six main stages as detailed below:

(i)   Establish the extent and nature of any rights to light that exist.
(ii) Establish the existing levels of sky visibility.
(iii) Establish the proposed levels of sky visibility.
(iv) Calculate the loss of sky visibility.

(v) Establish whether the anticipated loss of sky visibility merits any special enhancement factors.

(vi) Prepare the valuation.

Not all of the stages set out above will always be applicable but the starting point is always to establish whether or not rights exist and whether the particular rights are contractual or prescriptive in origin and permissive or restrictive in nature.

(i) Should the right to light prove to be contractual and restrictive, it is important to establish at an early stage whether or not the injured party is prepared to permit the development in a form that will be in breach of the restrictive covenant. It is also important to establish whether the breach will be of sufficient scale to merit an injunction. If the injured party is not prepared to permit the development in the form as designed and the breach is one that may well result in an injunction being granted, then the scheme may need to be redesigned to bring it within the permitted parameters. The difference in profit between the proposed development and the development permitted under the terms of any restrictive covenant is sometimes used to form the basis of any subsequently agreed settlement for releasing the covenant. In such a scenario a detailed Waldram diagram analysis is often unnecessary and the valuation comprises a development appraisal approach identifying the different profits achievable from the two options. The key issue from a practical consideration when dealing with a restrictive contractual right to light is to ensure that an early dialogue is established with the injured party.

(ii) Once the extent and nature of any rights to light have been identified, the next stage is to establish the existing level of sky visibility to those areas that will suffer a loss as a result of the proposed development. The method of calculating levels of sky visibility at a particular point in a room through a defined window aperture was established by Percy J Waldram. A Waldram diagram (figure 1) shows one half of the total dome of the sky on which is plotted the relevant window aperture and the various structures visible through that aperture. The diagram can then be used to calculate the amount of sky visibility at that particular point. Figure 2 shows a floor plan of a particular room with the relevant reference point (a) marked and figure 3 illustrates a Waldram diagram with the aperture and obstructions plotted, thereby showing the

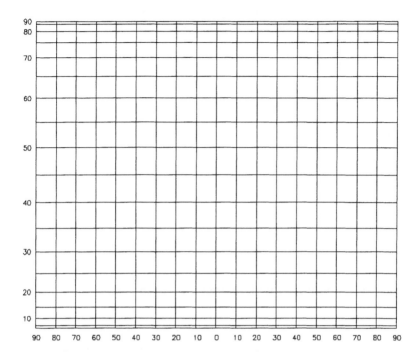

Figure 1. Waldram Diagram

existing level of sky visibility (Areas 2 and 3 combined) to that particular point. A Waldram diagram is then prepared for various points in the room to establish those points where 0.2% sky visibility exists. The resulting contour shows 0.2% sky visibility and enables the area of existing 0.2% sky visibility to be calculated.

(iii) The process is then repeated save that the proposed structure (Area 3) is plotted on the Waldram diagram in addition to the retained existing structures (figure 3). The points with 0.2% sky visibility will differ and a revised 0.2% sky visibility contour will be prepared reflecting the potential changes.

(iv) The area of loss of sky visibility can then be calculated by contrasting the area of existing 0.2% sky visibility with the area of proposed 0.2% sky visibility (figure 4). This basic process is then repeated for each room of the building which will suffer loss. The areas of loss are then assessed in EFZ terms and are

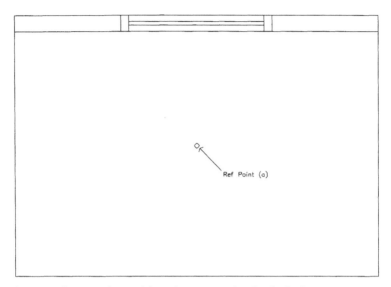

Figure 2. Room Plan with reference point included

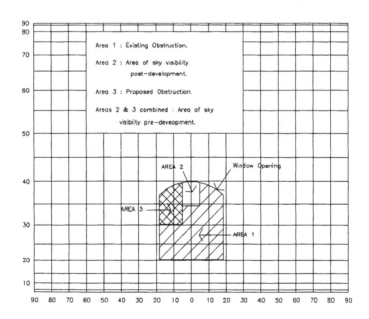

Figure 3. Waldram Diagram (for ref. point a) showing window in Figure 2, existing obstructions and proposed obstructions

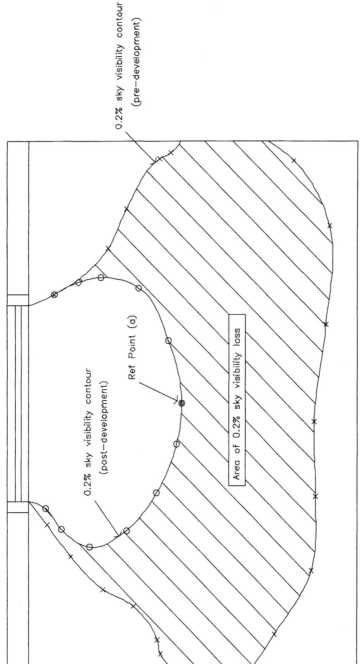

Figure 4. Room plan showing sky visibility contours

FIGURE 5
PROPERTY A - SCHEDULE OF LOSSES

FLOOR	ROOM/WINDOW	AREA	1/2 AREA	1/4 AREA	EXTG. 0.2% SKY	EXTG. 0.2% SKY %	PROP. 0.2% SKY	PROP. 0.2% SKY %	LOSS/GAIN	FRONT ZONE	FIRST ZONE	SECOND ZONE	MWT.	E.F.Z. TOTAL
Ground	G1	11.01	5.505	2.7525	9.89	89.83%	2.01	18.26%	7.88	0.7425	2.7525	2.7525	1.6325	5.650625
Ground	G2	4.96	2.48	1.24	4.94	99.60%	1.58	31.85%	3.36	0	0.9	1.24	1.22	1.825
Ground	G3	7.15	3.575	1.7875	6.72	93.99%	1.88	26.29%	4.84	0	1.695	1.7875	1.3575	2.928125
Ground	G4	10.76	5.38	2.69	10.14	94.24%	1.79	16.64%	8.35	0.9	2.69	2.69	2.07	5.9025
Ground	G5	10.69	5.345	2.6725	10.62	99.35%	3.04	28.44%	7.58	0	2.305	2.6725	2.6025	4.291875
First	F1	9.96	4.98	2.49	9.9	99.40%	3.1	31.12%	6.8	0	1.88	2.49	2.43	3.7325
First	F2	8.04	4.02	2.01	7.97	99.13%	2.23	27.74%	5.74	0	1.79	2.01	1.94	3.28
First	F3	11.31	5.655	2.8275	10.67	94.34%	3.26	28.82%	7.41	0	2.395	2.8275	2.1875	4.355625
First	F4	17.53	8.765	4.3825	15.32	87.39%	4.32	24.64%	11	0.0625	4.3825	4.3825	2.1725	7.210625
	TOTALS (ROUNDED) SQ.MT.								62.96	1.71	20.79	22.85	17.61	39.18

FIGURE 6
PROPERTY B - SCHEDULE OF LOSSES

FLOOR	ROOM/WINDOW	AREA	1/2 AREA	1/4 AREA	EXTG. 0.2% SKY	EXTG. 0.2% SKY %	PROP. 0.2% SKY	PROP. 0.2% SKY %	LOSS/GAIN	FRONT ZONE	FIRST ZONE	SECOND ZONE	MWT.	E.F.Z. TOTAL
Ground	G1	17	8.5	4.25	12.54	73.76%	1.35	7.94%	11.19	2.9	4.25	4.04	0	10.62
Ground	G2	13.23	6.615	3.3075	11.47	86.70%	2.1	15.87%	9.37	1.2075	3.3075	3.3075	1.5475	7.159375
Ground	G3	26.26	13.13	6.565	24.73	94.17%	5.68	21.63%	19.05	0.885	6.565	6.565	5.035	12.43375
First	F1	35.94	17.97	8.985	35.85	99.75%	7.12	19.81%	28.73	1.865	8.985	8.985	8.895	18.49875
First	F2	26.92	13.46	6.73	26.49	98.40%	7.98	29.64%	18.51	0	5.48	6.73	6.3	10.42
Second	S1	16.11	8.055	4.0275	15.65	97.14%	2.94	18.25%	12.71	1.0875	4.0275	4.0275	3.5675	8.564375
Second	S2	12.97	6.485	3.2425	9.61	74.09%	1.56	12.03%	8.05	1.6825	3.2425	3.125	0	7.32875
Second	S3	26.86	13.43	6.715	26.31	97.95%	5.625	20.94%	20.685	1.09	6.715	6.715	6.165	13.24875
	TOTALS (ROUNDED) SQ.MT.								128.30	10.72	42.57	43.50	31.51	88.27

typically presented in a schedule of losses such as those shown in figures 5 and 6.

(v) Once the losses of sky visibility have been calculated it is possible to establish whether or not the losses merit any special enhancement factors. From the schedule at figure 5 there is only a small area of front zone loss to property A and in those circumstances such a loss would not generally merit any special enhancement factor. The losses to property B show a far more significant front zone loss and these are likely to merit a special enhancement factor. If property B was residential then the extent of the front zone loss is such that a settlement based on a notional share of the development profit might be negotiated, assuming of course that the injured party is prepared to accept compensation and does not require the development to be redesigned.

(vi) In preparing the compensation valuations the first stage is to establish the rental values for each of the relevant buildings and also to establish the appropriate yield that would apply.

If the rental value for property A is £161.46 per m^2 (£15 per sq ft) the appropriate light rental value might be around £45 per m^2 (£4.18 per sq ft). Adopting an appropriate yield of 8%, we would then calculate the value of the infringement of the right to light as follows:

EFZ loss	39.18m^2
Agreed light rental value	£45.00 per m^2
Total Light Rental Value	£1,763.10 pa
YP perp at 8%	12.5
	£22,038.75
Say	£22,000

Consequently the appropriate level of compensation for the freeholder in possession of property A would be £22,000 (book value).

Should there be any tenants of property A who have rights to light, then this figure would be apportioned depending on either the length of the unexpired lease term or the period to the next rent review. For example, if a tenant had three years to the next rent review, the compensation payable to him would be based on the annual rental value capitalised for three years at 8%. The balance of the book value would be payable to the freeholder.

If similar rental and yield figures are applied to the losses for property B, the appropriate compensation payment (book value) would be, £49,650 as detailed below:

EFZ loss	88.27m^2
Light rental value	£45.00 per m^2
Total light rental value	£3,972.15
YP Perp at 8%	12.5
	£49,651.88
Say	£49,650.00

In the case of property B there is a significant area of front zone loss and consequently the parties might agree an overall enhancement factor of 2.5 in accordance with the principles set down in the *Carr-Saunders* case referred to above. Such an enhancement in this instance would give a compensation figure, after rounding, of £124,000. The amount of enhancement and the way it is calculated is a matter for negotiation. Depending on the particular circumstances it might be appropriate to enhance only the value attributed to the front zone loss. Alternatively if the loss is particularly severe a greater enhancement than 2.5 times might be applied to the total loss.

Chapter 6

Contaminated Land

The basic concepts underlying contaminated land

Background

The current contaminated land legislation came in section 57 of the Environment Act 1995, which retrospectively introduced a Part IIA into the Environmental Protection Act 1990. Although the legislation was put in place in 1995 it was not implemented in England until 1 April 2000.

One of the problems with contaminated land is that, to the person in the street, it is not necessarily clear exactly what contaminated land is. For something to be contaminated, in lay terms, it is not necessary for there to be danger or harm attached, although the *Concise Oxford Dictionary* (2002) does define contaminate as, 'make impure by exposure to or addition of a poisonous or polluting substance'. For the purposes of contaminated land, in the UK, contamination has been defined by the Environment Act 1995. This utilises a concept of harm which is considered later. For now, it is worth asking the question, 'where does land contamination come from?' A first reaction would probably be to suggest that land contamination arises from human activities such as industrial processes. This is, of course, true. However, there are natural sources of contamination as, for example, where radon gas is released from granite rock and contaminates the immediate vicinity. Similarly, where a stream from an underground source, contains higher than WHO[1] permitted levels of lead in drinking water. There are other sources of natural contamination, but the fact that contamination can arise naturally is an important point to be made. This text is, however, primarily concerned with contamination arising out of human activity, both past and present.

Readers will be aware that Britain went through an agricultural revolution and then, subsequently, an industrial revolution which led to Britain's prominence as a trading nation. Modern planning

---

[1]   World Health Organisation.

and health and safety controls were not always present. As a result, early industrial development had little or no control either as to where it took place or as to what activity took place. Nor were the storage of raw materials and disposal of waste materials subject to controls. The situation was superbly described by Charles Dickens when introducing his readers to the mythical Victorian urban development, 'Coketown' in *Hard Times*:

> It was a town of red brick, or of brick that would have been red if the smoke and ashes had allowed it; but as matters stood it was a town of unnatural red and black like the painted face of a savage. It was a town of machinery and tall chimneys, out of which interminable serpents of smoke trailed themselves for ever and ever, and never got uncoiled. It had a black canal in it, and a river that ran purple with ill smelling dye.

These, relatively uncontrolled, activities resulted in the pollution of land, water and air, many of which are still present today. Although the country's heavy industrial base may have declined and pollution control improved, Britain still has an active industrial sector and contamination of land, water and air can, and still does, occur.

One of the earliest, and probably the most notorious, cases of land contamination was that of Love Canal. This was a chemical dumping ground which developed, over time, into a 15-acre neighbourhood of the City of Niagara Falls, New York, USA. During the 1940s and 50s, the Hooker Chemical Company filled the canal, with about 21,000 tons of organic solvents, acids, and pesticides as well as their by-products. Dumping continued until the land was purchased for the construction of a school, despite warnings from the company. Eventually, during the 1970s, serious problems emerged. Children and animals received chemical burns and vegetation would not grow in certain areas. Epidemiological studies found birth defects, miscarriages, low birth-weight, cancers and respiratory disorders in people who lived in the area. In 1978, the then President of the United States, Jimmy Carter, declared the Love Canal area a federal emergency. Over 800 lawsuits were filed naming the Hooker Chemical Company, the City, County and Board of Education. This can be considered a seminal point in the history of land and contamination issues. Thereafter, the principle of liability for the adverse affects of land contamination, now known as 'polluter pays,' was established. This concept has largely been followed by UK contaminated land legislation, with a fall back position that, if all other potentially liable parties are incapable of providing sufficient redress, the landowner may be liable.

Legal framework

In order to appreciate the legal and financial impacts on contaminated land in the UK, a brief outline of the legislation is given. To fully appreciate the implications, it is recommended that a reputable legal authority and/or the statute is consulted. For the purposes of brevity, the scope and wider implications of the law have been summarised.

No definitive global view on what constitutes contaminated land currently exists and there are several definitions. One is provided by NATO:

> Land that contains substances that, when present in sufficient quantities or concentrations, are likely to cause harm, directly or indirectly, to man, the environment, or on occasions to other targets.

Notice the use of the term 'targets' in the above definition.

The concept of 'targets' and the spirit of the NATO definition would appear to have been incorporated into the UK legal concept of contaminated land.

Part IIA of The Environment Protection Act 1990, as amended by section 57 of The Environment Act 1995, and SI 2000, No 227, The Contaminated Land (England) Regulations 2000, provide the statutory framework for addressing contaminated land issues.

Contaminated land is defined in section 78(A) as

> ...any land which appears to the local authority in whose area it is situated to be in such a condition, by reason of substances in, on or under the land, that –
> (a) significant harm is being caused or there is a significant possibility of such harm being caused; or
> (b) pollution of controlled waters is being, or is likely to be, caused...

Substance

A substance is taken by section 78 A (9) to mean any natural or artificial substance, whether in solid or liquid form or in the form of a gas or vapour...

Pollution of controlled waters is defined as

> ...the entry into controlled waters of any poisonous, noxious or polluting matter or any solid waste matter

– Section 78 A(9).

For England and Wales, 'Controlled Waters' has the same meaning as in Part III of the Water Resources Act 1991– section 78 A (9) and, broadly, are:

1. defined areas of coastal waters,
2. inland freshwaters, and
3. groundwaters.

Harm means

By virtue of section 78 A (4) is taken to mean:

> harm to the health of living organisms or other interference with the ecological systems of which they form part and, in the case of man, includes harm to his property...

'Significant harm' will be taken as:

(a) Death, serious injury, cancer, other diseases, birth defects, genetic mutation or reproductive impairment of humans.

(b) An irreversible or adverse change to any living organism or eco-system in a protected habitat, eg a Site of Special Scientific Interest (SSSI).

(c) Structural failure or substantial damage of a building, excluding plant and machinery, which renders all or part of it unsuitable for its intended use.

(d) Death, disease or other physical damage involving a substantial loss in value of livestock, commercial game or crops, where they are rendered unsuitable for their intended use.

Thus, in order to be contaminated, land must be capable of causing harm.

Local authorities are required to inspect their areas in order to identify such land. In accordance with the guidance issued by the Secretary of State, they must determine whether 'harm' is to be regarded as 'significant', and whether the 'possibility' of significant harm being caused is 'significant'. They must also consider whether pollution of controlled waters is being, or is likely to be caused – section 78 A(5). If, in their opinion, the land is found to be contaminated, the authority must notify the Environment Agency, the landowner and any other person having responsibility.

The worst contaminated land is classified as 'special' and comes under the direct control of the Environment Agency. A detailed description of land, which is to be determined as a Special Site, is contained within Schedule 1 to the Contaminated Land (England) Regulations 2000.

In other cases, the enforcing authority is the district council or, in unitary areas, the appropriate local authority. The appropriate enforcing authority must serve a remediation notice on the 'appropriate person' specifying what that person has to do to minimise environmental damage.

Appropriate person

By virtue of Section 78(F), an appropriate person is any person, or any of the persons, who caused or knowingly permitted the substances, or any of the substances, by reason of which the contaminated land in question is such land to be in, on or under that land.

Remediation

Remediation is covered by section 78A(7) and is the doing of anything for the purpose of assessing the condition of the contaminated land in question, controlled waters affected by such land: or any land adjacent to such land.

It is probable that the authority will serve a remediation declaration. This will be appropriate where there is nothing that a charging notice could usefully require the recipient to do, steps are in hand for remedying the contamination or the authority already has power to carry out the remediation work. The declaration should indicate why the notice has not been served.

There is an appeal procedure, against the notice, to a magistrates court or the Secretary of State. If the notice was served by the Environment Agency, an appeal can be made to the Secretary of State for the Environment, Transport and the Regions. Non compliance with the notice can result in criminal penalties. If prosecution in the magistrates court is likely to be ineffective, an application can be made for an injunction in the High Court.

Non compliance can also result in the use of default powers by the enforcing authority. Outstanding costs may be charged on the land to which it relates but only if the landowner actually knowingly permitted or caused the contamination to occur. To charge land with outstanding remediation costs, an authority must first serve a charging notice on the person responsible. The consequence will be that, if payment is not made, the authority may sell the land and recover its costs. Again, there are full rights of appeal this time to the county court. Once the Act is in force, each enforcing authority will be required to maintain a public register of remediation notices, statements and declarations.

Liability

Liability falls to a Class A or a Class B person. In the first instance, the Class A person will be liable. If the class A person cannot be found, or is incapable of paying, then liability consequently falls to the class B person.

Section 78F(2) and (3) defines a Class A person as

> any person, or any of the persons, who caused or knowingly permitted the substances, or any of the substances, by reason of which the contaminated land in question is such land to be in, on or under that land.

A Class B person comes into play where, after reasonable inquiry, no person has been found to bear responsibility as an appropriate person in relation to things which are to be done by way of remediation. In this case, the owner or occupier for the time being of the contaminated land is an appropriate person. (Section 78 F(4) and (5)).

The owner is a person (other than a mortgagee not in possession) who, whether in his own right or as a trustee for any other person, is (or would be) entitled to receive the rack rent of the land. (Section 78 A(9)).

A person acting in relevant capacity: includes insolvency practitioner, official receiver, special manager, as defined by the Insolvency Act 1986. Such a person shall not be personally liable to bear the whole or any part of the cost of doing anything by way of remediation, unless that thing is to any extent referable to substances whose presence in, on or under the contaminated land in question is as a result of any act done or omission made by him which it was unreasonable for a person acting in that capacity to do or make – Section 78 X(3).

Contaminated land concept

What constitutes contaminated land, in 'lay terms', is not necessarily the same as that which would be considered contaminated in legal terms. To more fully appreciate the nuances, the UK definition of contaminated land will now be considered in more detail. In order for land to be considered 'contaminated', it must appear to the 'appropriate authority' to have a 'significant risk of causing significant harm'. It may well be that different 'appropriate authorities' will have different views on what is considered contaminated.

In order to be capable of causing significant harm there are three aspects which must be considered. There must be:

1. A contaminating substance (capable of causing harm),
2. A receptor or target (capable of being harmed), and
3. A pathway (the means by which the harmful contaminant comes into contact with and harms the receptor or target).

Contaminating substance

The contaminating substance is one which must be capable of causing significant harm. If the substance is relatively benign, then it will not be considered to be 'contaminating' and, so long as there are no other contaminating substances which are harmful, the land will not be classified as contaminated. The consequence of this definition is that the concentration of a substance must be such that it has a significant possibility of causing significant harm. Thus the legislation is not only concerned with the actual substance but also its level of concentration. This must be sufficient to pose a significant possibility of causing significant harm. One implication of this concept is, that where land has a contaminant present at levels which do not give rise to a significant possibility of significant harm, it may be considered to be polluted but not contaminated.

Receptor or target

In order for a substance to have a significant possibility of causing significant harm, there must be something for it to harm. Among other things, this might be a watercourse, structure or person. These 'objects' or 'things' can be considered as the receptor or target. If there is no receptor then it is possible to pose the question,

'Is the contaminant capable of causing significant harm?' and 'To what could any harm be caused?' If there is no target present then the answer must be 'nothing'. Consequently, the land would not be considered to be contaminated.

Pathway

In order to have a significant chance of causing significant harm, there has to be a way for the contaminating substance to reach the receptor. This is termed a pathway. Pathways can be varied. Their relevance depends upon the nature of the contaminating substance and the potential targets. Among the ways in which contaminants can reach targets are, inhalation and ingestion. If a contaminant reaches the surface and becomes an airborne dust, then inhalation may result in significant harm. An example might be asbestos fibres, which have been buried in a shallow excavation. During warm dry weather, surface dust may contain asbestos fibres. These may then be inhaled by humans. The resulting exposure will have increased their risk of developing lung cancer.

Another possible pathway would be illustrated by the presence of old lead paint dust buried in garden soil. This might be ingested by very young children, who tend to explore and put various substances in their mouths. Young children are particularly sensitive to lead poisoning, which has the potential to accumulate over a lifetimes' exposure.

A further pathway could comprise soluble toxic substances leaking (leeching) into a watercourse, which is subsequently, a source of drinking water. Humans, animals or eco-systems subsequently drinking or absorbing the water could be adversely affected and may suffer significant harm.

In principle, this is the UK legal concept of contaminated land, that is, land where substances are in such concentrations, where there are appropriate receptors and where there are pathways such that there is a significant possibility of significant harm. It follows that if any of these are absent, the land may appear to be uncontaminated for the purposes of a strictly legal definition. That is not to say it is not 'polluted' but, that in the UK, it is not considered to be contaminated. It may be that the lay person does not take such a view. It is also possible that substances, which are currently considered safe in themselves or at their existing concentrations are, at some time in the future, reclassified as unsafe. This might result from increases in safety standards or advances in scientific knowledge relating to adverse health effects arising from contaminating substances.

Contaminated land investigations

In order to be able to identify whether a contaminating substance offers the possibility of significant harm, it is necessary to understand which chemicals and uses have this potential. Fortunately, determining which chemicals are safe and which are not is, in principle, not that difficult. A number of sources exist, both on the Internet and in paper form. Consultation with these lists will help formulate a picture of what problems are presented. See identification of contaminants, below and the recommended information sources at the end of the chapter. While a pollutant may have the opportunity to cause harm, as shown earlier, it is essential to understand the potential pathways as well as appreciating the concentrations at which the possibility for harm becomes significant. It is not a decision that the valuer is called upon to make, but an appreciation of which chemicals are safe, and which are not, is informative.

Identification of contaminants

In order to identify the risks arising from contaminations on a site, it is not necessary for the soil to be investigated for the presence of *all possible* chemicals. This would be an exhaustive, extensive and expensive process. Only the harmful chemicals identified as being present need to be investigated in detail. Fortunately, most of these are well known as a result of past experience. The Environment Agency, has compiled industry profiles which can be readily purchased from the Department of the Environment, Transport and the Regions. These industry profiles list the typical pollutant(s) likely to be encountered for particular uses (see Stage I investigations below). Most of the commonly encountered polluting land uses are covered:

1. Airports.
2. Animal and animal products processing works.
3. Asbestos manufacturing works.
4. Ceramics, cement and asphalt manufacturing works.
5. Chemical works: coatings (paints and printing inks) manufacturing works.

6. Chemical works: cosmetics and toiletries manufacturing works.
7. Chemical works: disinfectants manufacturing works.
8. Chemical works: explosives, propellants and pyrotechnics manufacturing works.
9. Chemical works: fertiliser manufacturing works.
10. Chemical works: fine chemicals manufacturing works.
11. Chemical works: inorganic chemicals manufacturing works.
12. Chemical works: linoleum, vinyl and bitumen-based floor cover manufacturing.
13. Chemical works: mastics, sealants, adhesives and roofing manufacturing works.
14. Chemical works: organic chemicals manufacturing works.
15. Chemical works: pesticides manufacturing works.
16. Chemical works: pharmaceuticals manufacturing works.
17. Chemical works: rubber processing works (including works manufacturing tyres or other rubber products).
18. Chemical works: soap and detergent manufacturing works.
19. Dockyards and dockland.
20. Engineering works: aircraft manufacturing works.
21. Engineering works: electrical and electronic equipment manufacturing works (including works manufacturing equipment containing PCBs).
22. Engineering works: mechanical engineering ordnance works.
23. Engineering works: railway engineering works.
24. Engineering works: shipbuilding, repair ship-breaking (including naval shipyards).
25. Engineering works: vehicle manufacturing works.
26. Gasworks, coke works and other coal carbonisation plants.
27. Metal manufacturing, refining and finishing works: electroplating and other metal finishing works.
28. Metal manufacturing, refining and finishing works: iron and steelworks.
29. Metal manufacturing, refining and finishing works: lead works.
30. Metal manufacturing, refining and finishing works: non-ferrous metal works (excluding lead works).
31. Metal manufacturing, refining and finishing works: precious metal recovery works.
32. Oil refineries and bulk storage of crude oil and petroleum products.
33. Power stations (excluding nuclear power stations).
34. Pulp and paper manufacturing works.

35. Railway land.
36. Road vehicle fuelling, service and repair: garages and filling stations.
37. Road vehicle fuelling, service and repair: transport and haulage centres.
38. Sewage works and sewage farms.
39. Textile works and dye works.
40. Timber products manufacturing works.
41. Timber treatment works.
42. Waste recycling, treatment and disposal sites: drum and tank cleaning and recycling plants.
43. Waste recycling, treatment and disposal sites: hazardous waste treatment plants.
44. Waste recycling, treatment and disposal sites, landfills and other waste treatment or waste disposal sites.
45. Waste recycling, treatment and disposal sites: metal recycling sites.
46. Waste recycling, treatment and disposal sites: solvent recovery works.
47. Miscellaneous industries:
    - Charcoal works.
    - Dry-cleaners.
    - Fibreglass and fibreglass resins manufacturing works.
    - Glass manufacturing works.
    - Photographic processing industry.
    - Printing and bookbinding works.

It should be pointed out that agricultural uses can also pollute land. Examples might be sheep dips, which have contained pesticides, pits used for burying animals infected by foot and mouth and storage tanks for diesel fuel used for machinery, and which have leaked over prolonged periods of time. This list is anything but exhaustive, but it does indicate the type and nature of the potential problem. Nor should it be assumed that a site is uncontaminated simply because it is 'green', or previous development is not apparent. For example, a valley site may appear green, yet be subject to pollutants migrating with rain water from long abandoned spoil tips on the surrounding hills. The industrial uses which had produced the polluting spoil heaps having ceased and the spoil tips naturally vegetated over, the overall appearance is that of a green valley set within rolling green hills. Even so, the historical legacy of contamination at the base of the valley may be extremely high! It is important to appreciate that it is possible for contaminants to migrate onto or off a site. The lesson is that neither a lack of apparent industrial processes nor a green appearance are guarantees of non contamination.

In order to commence identifying potentially polluting chemicals, it is helpful to first identify current and previous uses of the land. (Remember that adjoining and nearby uses can also contribute to site contamination.) This can be done in a number of ways including the use old maps and local memory. It is not the intention to discuss these methods in detail.

Important caveat

This work is primarily aimed at an appreciation of the problems of valuing contaminated land. The following sections on investigations have been produced to enable an appreciation of the context of the valuation process. In practice, no investigations should be undertaken without appropriate education and training to ensure competence. It is also necessary to have appropriate and sufficient professional indemnity insurance cover in place.

Site investigations

When a site is presented to the valuer, it may or may not be evident as to whether there is any contamination on the site. It could be that the site is covered with industrial buildings, that there are no signs of industrial buildings, that the site is covered with some other use or appears to be a green field site. As a consequence, it is not possible to be certain as to the position regarding contamination without undertaking some form of investigative process.

Land has to be valued, therefore, either on the assumption that there is no contamination (with the proviso that should contamination be found the valuation will be revisited) or with adequate knowledge of the contamination and its implications. In order to undertake a valuation on the latter basis, the valuer should have a good knowledge of the nature, extent and remediation costs of the contamination. This will almost certainly require the use of appropriately qualified external consultants who should also carry appropriate and adequate levels of professional indemnity insurance.

There are several environmental consulting firms that offer this sort of advice. They should be chosen with care, having regard to their competence and experience, in the light of the anticipated site contaminations. It should also be remembered, that in the event of a claim, a consultant's advice may only be as good as the level and standing of their professional indemnity insurance. Unless care is taken, surveyors could unwittingly find themselves joined in a claim for negligence against the consultant. It is often advisable for the client to be responsible for the appointment of a consultant, in order to ensure that the valuer does not assume undue liability.

Stage I (Phase I) and Stage II (Phase II) investigations

The first part of any investigation would revolve around inquiries into readily available and accessible information. This type of investigation is usually referred to as *stage one or phase one* investigation. The main purpose is to identify current and previous uses of the site and its immediate surroundings. A phase one investigation is useful in determining whether a site is likely to be contaminated, the anticipated nature and probable extent of contamination. Investigations may go back many years and can reveal that a site has had a variety of uses, occupiers and may have been exposed to several contaminating activities over its life. The results of a phase one inspection help when contracting out phase two investigations.

It may be that the vendors or the proposed purchasers have already commissioned a report into the levels of contamination. Any such report should be treated with care, having regard to the laws of contract and liability. The valuer may not be covered when relying on such information. Investigations would normally proceed along the lines of identifying current and previous uses on the site. Adjoining sites would usually be included where there is the potential for on/off site contamination.

Initial information may come from Internet based searches with further evidence coming from:

- on site inspection,
- records and plans of the existing owner/occupants,
- old plans maps and records held in the local library/local authority,
- living memory and any other similar inquiries, and
- specialist search companies.

Phase two or stage two investigations are usually the main way of obtaining detailed information. They are usually intrusive and comprise a more detailed inspection and testing of the site. They would be expected to include sampling, other ground investigations and maybe the use of boreholes. Analysis of samples may result in further and more specific investigations being required.

It is essential that such investigations and sampling should be undertaken following good practice, if the results are to be reliable. The findings can have a major impact upon the:

- treatment,
- remediation costs,
- potential liability,
- potential for valuation stigma, and
- nature of the redevelopment proposals for the site.

When the investigations have been completed, and the type, nature and extent of contaminations are known, it is possible to determine what types of treatment are required. These may also depend upon the proposed end use and other factors.

Treatment

As mentioned earlier, contaminated land is considered to be contaminated because there is a contaminant, pathway and receptor/target and the contaminant must have the significant possibility of causing significant harm. If any of these aspects are removed, then it is possible for a contaminant, within an area of land, to be considered non-contaminating, even though, in reality, it may still exist. First the pollutant will be considered. The pollutant must exist in sufficient concentrations for it to be able to cause the possibility of significant harm. Its potential for producing significant harm might be reduced if the concentration of the chemical or pollutant is reduced to such an extent that such a possibility no longer exists. This can be considered as reducing the level of the pollutant below the trigger or threshold level. The trigger or threshold level is the level at which the possibility of causing harm becomes significant, having regard to the proposed end use of the land. This is called the *'suitable for use'* approach to treatment. The other main approach is that of *'multiple use'*, normally total clean up. This is commonly used in countries such as Holland and the USA. The multiple use approach requires a high standard of clean up which normally allows the land to be used for most purposes. Consequently, it may be more expensive than the suitable for use approach.

An advantage of the suitable for use approach, is that it is pragmatic. The land is treated in terms of the proposed end use. Consequently, the standard of remediation may be expected to be greater for residential purposes than say that for car parking. Furthermore, the nature of the development (as is discussed later) may, itself, decontaminate the site (in the context of the UK legal definition). The resulting effect is that land may be decontaminated (in legal terms) as a consequence on a suitable for use basis, where a multifunctional (multiple use) approach would have been financially prohibitive. In this way, although the suitable for use approach may seem the lesser standard, the risks arising from contaminated land are reduced more quickly.

As mentioned earlier, in order for the pollutant to reach the target, there must be a pathway. It follows that if no pathway exists, or an existing pathway is removed, then for the purposes of UK legislation, the land will no longer be considered contaminated. This is true even when the contaminant actually remains on site. A number of means for achieving this may be undertaken. They all have the same function and that is the contaminant is isolated from potential targets. In simple terms this may mean that the contaminated element of land is encapsulated. Encapsulating the contaminant ensures that it can no longer reach any target.

In order to ensure that this remains effective, it may be essential to continue to monitor the site into the foreseeable future. It is also the case that where existing contaminants remain on site they may have to be reconsidered at some future date when the site is redeveloped.

Often the simplest ways, and certainly the most common, to remediate contaminated land, is to remove the contamination from the site. This is commonly known as 'dig and dump'. Unlike the other methods of treatment, this has the advantage that the 'remediated' site is known to be clean. Consequently, it is known as a multi-functional approach to clean-up because the decontaminated site is suitable for many uses, subject to normal planning constraints. Pollutants may also be removed or reduced to acceptable levels by chemical or biological remediation techniques.

Development and redevelopment

A proposed redevelopment may create a target for existing contamination. The threat may be removed either by removing the contaminant, treating the contaminant to reduce the risk or removing the pathway. As discussed earlier, the significant possibility for significant harm may also be affected to a degree by the proposed use for the site. For example, a residential development, which would include gardens that might be used for growing vegetables, would normally require a greater degree of treatment than a proposed industrial use. It is also worth remembering that the nature of the development may affect the pathway. If a contaminant were naturally encapsulated within the ground and the redevelopment resulted in an impermeable capping, such as a concrete raft, then the pathway would be removed by virtue of the redevelopment itself. Consequently the specific requirements of an actual development proposal would have to be considered in detail.

Any contamination left on the site will usually require monitoring into the foreseeable future, to ensure that contamination does not escape. It may have the potential to impinge on any use subsequent to the proposed development. This might not at first sight appear to be a problem in the UK, where the functional and economic life-span of property are commonly both long. It may be however, that the economic life is less than the functional life, and early redevelopment may even be anticipated. Under such circumstances, whether to totally remove contamination or encapsulate it may be a particularly pertinent factor for consideration.

Funding bodies, insurers, investors and purchasers may also have concerns about the nature of any residual contamination and its potential ability to stigmatise a development. That is not necessarily to say that all redeveloped contaminated sites must be stigmatised, but rather to assert that the potential for stigma exists and should be considered and quantified where possible. The nature of treatment and any residual contamination may have an effect upon stigma. Even where land is decontaminated and the owner is indemnified against future contamination, a drop in value has been identified called stigma.

The valuation problems of remediation and uncertainty over financial costs which result in stigma have been considered by Patchin, (1994). He defined stigma as, 'The value impact of environmentally-related uncertainties' (Patchin, 1988, 1992 and 1994). According to Collangelo and Miller stigma is, 'The negative impact that results from public perceptions that environmental contamination is permanent and represents a continuing risk even after environmental clean up has been completed...' (Collangelo and Miller, 1995). In the UK, the 1996 edition of the 'Red Book' suggested that the collective factors influencing stigma were:

(a) inability to effect a total cure;
(b) prejudice arising out of past use(s);
(c) the risk of failure of treatment;
(d) compensation payable or receivable under section 78(G) of the Environment Protection Act 1990 or otherwise;
(e) the risk of change to legislation or remediation standards;
(f) a reduced range of alternative uses of the site; and
(g) uncertainty.

(GN 2.9.7 RICS 1996)

As mentioned earlier, the proposed end use of the land will have an influence upon the nature of the treatment. It is probable that residential, commercial and industrial redevelopments will have their own characteristics for consideration. Some elements of the risk attributable to contamination can be insured against with specific insurance policies. These are normally instituted on a claims made basis, ie the insurance cover exists for the specific year when a policy is in existence.

Possible advantages of contaminated land

One point that is worth mentioning, is that contaminated land can offer some positive advantages in redevelopment. Planning permission may be eased by the potential for a contaminated and derelict site being returned to a beneficial and attractive use which may also enhance the facilities in a region and create employment.

It is axiomatic in valuation that return is proportional to risk, ie the greater the risk the greater the return. This perhaps should be rewritten as, the greater the perceived risk the greater the return. Where a risk is perceived by the market as being great, but experience suggests that the risk is less than the market perception, potential for super profits may exist. Understanding the risks involved in development reduces the actual risk to which a developer is exposed. A developer who understands the risk better than another is more likely to maximise profits from development.

Where a 'green site', or a site where there is no expected contamination is being considered, an element of uncertainty must remain. It may be possible that the site is contaminated. The only way of being certain would be to commission an investigation of the site. By comparison, a formerly contaminated site, which has been treated in accordance with the appropriate standards and by competent professionals, is a known risk. A case could be made, that in certain circumstances, this certainty might even carry a premium, with the site being worth proportionally more than an apparently 'green' site, where the possibility of contamination is an unknown.

Professional requirements

It is absolutely essential to comply with professional guidelines and to have the appropriate expertise and professional indemnity insurance cover. Not complying with guidelines or being unaware of them, could be construed as negligence. In the UK, the valuer should be familiar with the appropriate RICS publications.

At the time of writing, the 'Red Book' has been rewritten but the sections dealing with contaminated land have not yet been revised. For the time being, it would be wise to remind all surveyors to ensure that they are familiar with the relevant sections of the 'old' Red Book, in particular PS 6.3, GN 2.7 and GN 2.9. The surveyor should read and be familiar with the actual publication. For the purposes of this text, reference is made to some of the sections but not all of them. It is also worth mentioning that as stated in GN 2.10, '...you should investigate the adequacy of your professional indemnity insurance (PII).' While a valuer should not express an opinion on matters for which they have neither the PII cover, nor the expertise, failure to report potential contamination, which would be apparent to a competent valuer in the course of the provision of the valuation service, is considered to be a breech of the duty of care.

Taking instructions

It is important to clarify instructions at the outset – what is covered and what is not covered before a commission is accepted. The 'Red Book' makes the following points.

PS 6.3.3

> Confirm instructions at the outset in writing; be aware of your limitations and competence: Work as a team with lawyers and other specialist advisers. Where the extent of necessary investigations is found to be greater than had been anticipated at the time that instructions were agreed, the valuer must consult with the Client and agree amendments to the conditions of engagement before incurring additional costs.

(RICS 1998)

PS 6.3.5

> If there is nothing to indicate to the valuer that there is contamination then the property can be valued on that basis with suitable caveats.
>
> (RICS 1998)

PS 6.3.6

> If evidence is available to the valuer of contamination and the cost of rectification has been estimated by experts with appropriate competence, then having regard to the relevant use(s) being assumed, this can be reflected in the valuation, taking due account of any assumptions, the likely effect of statute law and perception of the market and a suitable caveat ... adopted.
>
> (RICS 1998)

PS 6.3.7

> If there is evidence of contamination, but its extent cannot be established for reasons such as absence of technical skills, time available, or costs, the valuer may: (a) decline the instruction; or (b) negotiate an acceptable basis for undertaking the work. [Here the practice statement goes on to suggest what these might be.]
>
> (RICS 1998)

Appointing specialist/consultants

As mentioned earlier, valuers should check their professional indemnity cover. Some policies exclude liability for negligence claims arising from contamination. Ensure that any specialists/consultants employed are appropriately qualified and also have appropriate professional indemnity cover. It might be prudent to ensure that clients appoint any specialists/consultants themselves, so that a contract exists directly between the client and the specialist/consultant. This helps to minimise any danger of the valuer being joined with the specialist/consultant in the event of a negligence.

Valuation theory

In principle, the valuation format is similar to that commonly undertaken by valuers where some form of impediment to land or property exists.

	Value of property in good condition (U)
Less	Costs of rectifying impediment (Cre)
Leaves	Amount available for purchasing the impaired property (I) ie The value of the property in its original impaired condition to a purchaser with no special interests in the property (OMV).

It should be remembered that a client purchasing a property on this simple equation might be better buying a property which is already in good condition. Why should they rectify the impediment if there is no financial gain? Consequently, where the return required for undertaking such an improvement is P, the equation can be rewritten:

$$OMV = U - (Cre + P)$$

This is still too simplistic. There is general agreement that any valuation should consider stigma. Stigma has been taken into account even in the USA where 'total clean up' has been undertaken. According to the 'old' Red Book, GN 2.9.7 suggests that the collective factors influencing stigma are:

(a) inability to effect a total cure;
(b) prejudice arising out of past use(s):
(c) the risk of failure after treatment
(d) compensation payable or receivable under section 78G of the Environmental Protection Act 1990 or otherwise;
(e) the risk of change to legislation or remediation standards;
(f) a reduced range of alternative uses of the site; and
(g) uncertainty.

Stigma may be factored into a valuation in a number of ways. A lump sum may be deducted or the capitalisation rate applied to the income stream may be adjusted. Whichever option is adopted it is important that the valuer can support the adjustments by reference to evidence. Such a valuation might then be written:

$$I = U - Cre - S$$

Where
I   = Impaired Value of land.
U   = Value of land in good condition (i.e. unimpaired).
Cre   = Costs of remediating the land to the appropriate standard.
S   = Deduction for stigma.

Wilson (1993) defined stigma as 'the value impact of uncertainties'. According to him, if written in more detail with the terminology used so far, the formula would be:

$$I = U - CNCP - CR - CF - S$$

Where
I   = Impaired Value of land.

U   = Unimpaired value (ie the value considering all restrictions on use and costs of ownership, other than environmental risk).

CNCP = Cost of restrictions on use and/or environmental liability prevention.

CRe = Costs of remediating the land to the appropriate standard.

CF   = Impaired financing costs.

S   = Stigma, or negative intangible cost.

(Wilson et al 1993)

Reduction for stigma

Syms (1996) has suggested that one way forward would be to make a percentage deduction from a valuation similar to that above. The following example is based on Syms suggestion.

	£	£
Unimpaired value of land		1 600 000
Remediation costs in accordance with GN 2 para GN 2.9.2		
clean up of on-site contamination	360 000	
effective contamination control and management	85 000	
redesign of production facilities	N/A	
penalties and civil liabilities for non compliance	N/A	
indemnity insurance for the future	11 000	
avoidance of off site migration to other site	90 000	
control of migration from other sites	12 000	
regular site monitoring	9 000	
Estimated total treatment costs		567 000
Anticipated economic life of buildings say 20 years		
Present value of £ for 20 years at 7.5%		0.235413
Present value of treatment costs		133 479
Value in the impaired state, excluding stigma		1 466 521

This is a reduction of 8.34%. Syms then makes a deduction from this figure to reflect stigma. The figure comes from analysis of a range of similar contaminated comparables for which there is reliable evidence. On the basis that such comparables show a loss of say 23–27% for stigma, a deduction of 25% may be applied in this case, thus:

	£
Value in the impaired state, excluding stigma	1 466 521
Less 25% to reflect stigma	366 630
Impaired value taking account of treatment, costs and stigma	1 099 890

This method has been commented upon by Wilbourn and Dixon, who suggested that a straight percentage deduction could be misleading. Dixon suggested that applying a fixed percentage deduction resulted in a high value property experiencing greater stigma than a low value property. Wilbourn was of the opinion, that in his experience, '...it is not possible to define a level of stigma discount allied to the level of contamination.'

Richards (1995) found another approach in his research, that of increasing the yield on the income stream to reflect the increased risk:

	£
Net income	125 000
YP in perpetuity at 12%[2]	8.33
	1 041 250
Less	
Net present value of remediation costs	250 000
Value allowing for treatment and stigma	791 250

Valuation case studies

The following case studies have been included to demonstrate how various aspects of contaminated land may be considered in practice. They cover a range of examples and are typical of those that may be encountered in practise.

Case study 1

A freehold industrial site is for sale on the open market. A valuation is carried out in the normal manner, as initially, it is believed that the site is unaffected by contamination. On the assumption that the site is not contaminated, the open market value is determined as being represented by a figure of £1,500,000.

Case study 2

Desk top investigations are now conducted into the earlier uses of the freehold site in case study 1. It is now found out there is a strong

---

[2]   This would normally have been 11% but has been increased by 1% to reflect stigma. Unless there is comparable evidence to support this, such alterations are very subjective and may be difficult to justify.

probability that the site has been contaminated with hydrocarbons. A consultants' report is commissioned and the contamination is identified as being within a clearly delineated area. Remediation costs are estimated at £300,000. The revised valuation will be:

		£
Value on an unimpaired basis		1 500 000
Cost of remediation	300 000	
Fees at say 10%	30 000	
Indemnity insurance	20 000	350 000
Value in contaminated condition		1 150 000

Contaminated land does not necessarily take the form of polluted development sites. The next three case studies deal with residential properties in established residential areas.

Case study 3

A 'right to buy' house is purchased, at an open market value of £38,000 in an area which had been subject to coal mining activity. It is found to be suffering from the ingress of gasses arising from the mines. The claimants' valuer contests that the property has a negative value. Eventually, the property is valued, with the contamination, at 40% of the open market value, ie £15,200.

This is agreed because it is decided that a public authority will rectify the problem by way of gas management and a new cap on the mine shaft, which is located in the garden. It is believed that there will always be a cash buyer for properties such as this. The hope value, associated with the purchase of a property at a substantial discount, cannot be ignored.

Case study 4

A portfolio of residential properties is to be acquired by an investment company. Many of the properties are found to be contaminated with mine gas, arising from the past mining uses, in locations where contamination could affect them. The portfolio is acquired at a 37% discount, ie 63% of the estimated open market value in good condition. The 37% deduction is made to reflect the uncertainty and the ability to break up and sell off the portfolio.

Case study 5

A residential property is contaminated by fuel oils leaking from a fuel oil storage tank, higher up the hill. The contamination is so great, that people can smell odours both outside and inside the house. This situation is aggravated by the fact that the water table is only 0.6 metres below the ground-floor level.

This is a complex case. Given the contamination, the property would be valued as a total loss. However, if substantial remedial works are undertaken, the property could be restored. The assumption is that the value of the property would then rise over a period of 10 years, back to open market value. This stepped approach reflects the fact that the stigma is greatest in the earliest years after remedial works are undertaken. The life of the stigma should decay similarly to that of a modern development site, which has a 10-year NHBC agreement with an element of environmental insurance. This is accepted by both parties. Consequently, after remedial works, the property is valued as having a current market value of £225,000.

Case study 6

A dental surgery investment property is to be constructed on what has been found to be a site contaminated with gas works waste. Remedial works have been estimated at £200,000 but the investment will only be worth £200,000. Consequently, there would appear to be no value in the property. The tax status of potential financing vehicles and availability of possible grants are explored. It transpires that no grants are available, nor can any advantage be obtained from differing funding options. As a consequence, the scheme will not go ahead. The proposed use for the site is, apparently, the most valuable so it is unlikely that the site will be developed. The result is urban blight.

Case study 7

This case study deals with stigma and assesses a deduction to be made for the risk of future contamination. It also concludes that, in valuation and stigma terms, contamination of part of a site may affect the value of the whole. It is of particularly interest, as it results from litigation which ended in the Court of Appeal. Consequently it has particular authority.

In the case of *Blue Circle Cement plc* v *Ministry of Defence* [1999] Env LR 22, radioactive matter, including plutonium, escaped from an adjacent site owned by the MOD. The contamination only contaminated part of the estate. The MOD subsequently undertook remedial works.

Their Lordships held, that even though a small part of the estate had been contaminated, the resulting stigma had impinged on the whole estate. It was decided that the site had experienced a depreciation of 25% arising from the contamination. It was also determined that the market value of the site was depreciated by a sum of 10% to reflect the risk of future contamination.

Case study 8

While not necessarily contaminated, landfill sites can have an effect on value. Recent research has been published which sheds light on how proximity to the actual presence of a landfill site can impact upon value. This is considered in the following hypothetical case study.

A detached property is situated in the East Midlands, a quarter of a mile from a landfill site. The value has been depreciated by the presence of that site. The valuation has drawn on properties in the immediate area and so no specific deduction for the proximity of the landfill site was made. However, during periods when there is a dearth of comparable evidence, it may aid interpretation of market data to be aware of research published by DEFRA (2003). This identified a reduction in property values attributable to their proximity to landfill sites. In the case of the East Midlands, this was found to be a 10.1 % deduction for properties situated within a quarter of a mile from a landfill site. The percentage deduction reduces as the distance from the landfill site increases. It also depends upon the regional location and, to some extent, on the nature of the materials used for fill. The greatest reductions were noted in Scotland, where values of properties within a quarter of a mile of a landfill site, could be depreciated by an amount in excess of 41%.

Further reading

DEFRA, (2003) *A Study to Estimate the Disamenity Costs of Landfill in Great Britain*, http://www.defra.gov.uk/environment/waste/landfill/disamenity.htm

DETR Industry Profiles

1. Airports (ISBN 1 851 122 893).
2. Animal and animal products processing works (ISBN 1 851 122 389).
3. Asbestos manufacturing works (ISBN 1 851 122 311).
4. Ceramics, cement and asphalt manufacturing works (ISBN 1 851 122 907).
5. Chemical works: coatings (paints and printing inks) manufacturing works (ISBN 1 851 122 915).
6. Chemical works: cosmetics and toiletries manufacturing works (ISBN 1 851 122 923).
7. Chemical works: disinfectants manufacturing works (ISBN I 85 112293 1).
8. Chemical works: explosives, propellants and pyrotechnics manufacturing works (ISBN 1 851 122 370).

9. Chemical works: fertiliser manufacturing works (ISBN 1 851 122 893).
10. Chemical works: fine chemicals manufacturing works (ISBN 1 851 122 354).
11. Chemical works: inorganic chemicals manufacturing works (ISBN 1 851 122 958).
12. Chemical works: linoleum, vinyl and bitumen-based floor cover manufacturing works (ISBN 1 851 122 966).
13. Chemical works: mastics, sealants, adhesives and roofing manufacturing works (ISBN 1 851 122 966).
14. Chemical works: organic chemicals manufacturing works (ISBN 1 851 122 753).
15. Chemical works: pesticides manufacturing works (ISBN 1 851 122 745).
16. Chemical works: pharmaceuticals manufacturing works (ISBN 1 851 122 362).
17. Chemical works: rubber processing works (including works manufacturing tyres or other rubber products) (ISBN 1 851 122 346).
18. Chemical works: soap and detergent manufacturing works (ISBN 1 851 122 761).
19. Dockyards and dockland (ISBN 1 851 122 982).
20. Engineering works: aircraft manufacturing works (ISBN 1 851 122 990).
21. Engineering works: electrical and electronic equipment manufacturing works (including works manufacturing equipment containing PCBs) (ISBN 1 851 123 008).
22. Engineering works: mechanical engineering ordnance works (ISBN 1 851 122 338).
23. Engineering works: railway engineering works (ISBN 1 851 122 540).
24. Engineering works: shipbuilding, repair shipbreaking (including naval shipyards) (ISBN 1 851 122 77 X).
25. Engineering works: vehicle manufacturing works (ISBN 1 851 123 016).
26. Gasworks, coke works and other coal carbonisation plants (ISBN 1 851 122 32 X).
27. Metal manufacturing, refining and finishing works: electroplating and other metal finishing works (ISBN 1 851 122 788).
28. Metal manufacturing, refining and finishing works: iron and steelworks (ISBN 1 851 122 80 X).

29. Metal manufacturing, refining and finishing works: lead works (ISBN 1 851 122 303).
30. Metal manufacturing, refining and finishing works: non-ferrous metal works (excluding lead works) (ISBN 1 851 123 024).
31. Metal manufacturing, refining and finishing works: precious metal recovery works (ISBN 1 851 122 796).
32. Oil refineries and bulk storage of crude oil and petroleum products (ISBN 1 851 123 032).
33. Power stations (excluding nuclear power stations) (ISBN 851 122 818).
34. Pulp and paper manufacturing works (ISBN 1 851 123 040).
35. Railway land (ISBN 1 851 122 532).
36. Road vehicle fuelling, service and repair: garages and filling stations (ISBN 1 851 123 059).
37. Road vehicle fuelling, service and repair: transport and haulage centres (ISBN 1 851 123 067).
38. Sewage works and sewage farms (ISBN 1 851 122 826).
39. Textile works and dye works (ISBN 1 851 123 075).
40. Timber products manufacturing works (ISBN 1 851 123 083).
41. Timber treatment works (ISBN 1 851 122 834).
42. Waste recycling, treatment and disposal sites: drum and tank cleaning and recycling plants (ISBN 1 851 123 091).
43. Waste recycling, treatment and disposal sites: hazardous waste treatment plants (ISBN 1 851 123 105).
44. Waste recycling, treatment and disposal sites, landfills and other waste treatment or waste disposal sites (ISBN 1 851 112 113).
45. Waste recycling, treatment and disposal sites: metal recycling sites (ISBN 1 851 122 29 X).
46. Waste recycling, treatment and disposal sites: solvent recovery works (ISBN 1 851 123 121).
47. Miscellaneous industries (ISBN 1 851 123 13 X).

Selected Web pages

http://www.environment-agency.gov.uk/netregs/
Netregs, Environment Agency pages with useful information on contaminated site investigations and accessibility to on-line UK statute.

http://216.31.193.171/asp/1_introduction.asp
What's in my back yard, Environment Agency pages which allow environmental information to be accessed by post code.

http://physchem.ox.ac.uk/MSDS/
The Physical and Theoretical Chemistry Laboratory, Oxford University, Chemical and Other Safety Information

http://www.claire.co.uk/
Contaminated Land: Applications In Real Environments, a public/private partnership involving: Government policy makers, Regulators, Industry, Research organisations, Technology developers.

http://www.nicole.org/
The Network for Industrially Contaminated Land in Europe.

Selected legal cases

Blue Circle Industries plc v *Ministry of Defence* [1999] Env LR 22

Cambridge Water Company v *Eastern Counties Leather plc* [1994] 1 All ER 53, [1994] Env LR 105

Celtic Extraction Ltd (in liquidation), Re [2000] 2 WLR 991; [1999] 4 All ER 684, *sub nom Official Receiver as Liquidator of Celtic Extraction Ltd and Bluestone Chemicals* v *Environment Agency* [1999] 3 EGLR 21; [1999] 46 EG 187; [2000] Env LR 86

Empress Car Company (Abertillery) Ltd v *NRA* [1998] Env LR 396

Environment Agency v *Clark (Administrator of Rhondda Waste Disposal Ltd)* [2000] Env LR 600

Environment Agency v *Stout* [1998] 4 PLR 56; [1999] Env LR 407

Graham v *ReChem International Ltd* [1996] Env LR 158

Land Catch Ltd v *International Oil Pollution Compensation Fund* (Scottish Case) (1999) ENDS Report 293 p51

R v *Dovermoss* [1995] Env LR 258

Shanks & McEwan (Teeside) Ltd v *Environment Agency* [1997] Env LR 305

Yorkshire Water Services Ltd v *Sun Alliance & London Insurance plc* [1998] Env LR 204

Selected statutes

Environmental Protection Act 1990 – Parts I, II, III and Pollution Prevention and Control Act 1999 Part II A (inserted by Environment Act 1995)
Radioactive Substances Act 1993
Town and Country Planning Act 1990
Water Resources Act 1991

Selected regulations

Contaminated Land (England) Regulations 2000, SI 2000, No 227
Pollution Prevention and Control (England and Wales) Regulations
2000

Recommended reading

Cairney T and Hobson DM (eds) *Contaminated Land* (2nd ed), (1998)
E&FN Spon, ISBN 0 419 23090 4.

Dixon T (1995) *Lessons from America: Appraisal and Lender Liability
Issues in Contaminated Real Estate*, College of Estate Management.

Hester RE and Harrison RM (eds) *Contaminated Land and
Reclamation* (1997) Thomas Telford, ISBN 9 780727 725950.

Patchin PJ (1994) Contaminated Property and the Sales
Comparison Approach, *The Appraisal Journal* (July). pp 402–9.

Paul V (1995) *Bibliography of Case Studies on Contaminated Land:
Investigation, Remediation and Redevelopment* BRE.

RICS (1996) *Appraisal and Valuation Manual* (The Red Book).

RICS (1995) *Land Contamination Guidance for Chartered Surveyors*
RICS.

Richards T (1995) *A Changing Landscape: The Valuation of
Contaminated Land and Property* College of Estate Management.

Syms P (1997) *Contaminated Land (The Practice and Economics of
Redevelopment)* ISBN 0 632 04134 X.

Syms P (1996) Dealing with Contaminated Assets, *Estates Gazette*
(23/3/96) pp 124–5.

Syms P *Desk Reference Guide to Potentially Contaminative Land Uses*
(ISVA) RICS Books, London, ISBN 0 9029 1303 4.

Chapter 7

Council Tax

Introduction

Council Tax was introduced by the Local Government Finance Act 1992 (the 1992 Act) to replace the previous Community Charge legislation that itself superseded the old system of domestic rating from 1 April 1990. Unlike Community Charge, which was a tax levied directly upon each individual taxpayer, or domestic rates that were based on the rental value of the property, council tax is calculated by reference to the capital value of a dwelling, that enables it to be placed in one of 8 valuation bands A to H which vary between Scotland, Wales and England.

In England and Wales responsibility for compiling and maintaining a list of valuation bands for each dwelling in a billing authority area is given to a statutory officer known as the Listing Officer, who is appointed by the Commissioners of Inland Revenue, but is part of the Valuation Office Agency. This list is known as the 'valuation list' and the current lists came into effect on 1 April 1993. In Scotland the responsibility is exercised by the local Assessor for each Regional and Islands council. The first general revaluation in Wales is to come into force on 1 April 2005 and in England on 1 April 2007. Thereafter it is proposed that there will be revaluations every 10 years.

The rules for identification of the dwelling to be valued, the valuation assumptions to be adopted, and the procedural rules for alteration of valuation lists and appeals are contained in the 1992 Act and a series of subsequent Statutory Instruments. Much of the council tax legislation has its roots in the rating system, and therefore case law relating to rating cases is often equally applicable to council tax.

Liability to pay council tax shall be determined on a daily basis, and the tax is collected by the local authority (billing authorities in England and Wales). This Chapter focuses on valuation matters, does not deal with aspects of collection, reliefs or exemptions, and only briefly refers to the regulations relating to appeals.

Definition of a dwelling for council tax purposes

Council tax shall be payable in respect of any dwelling which is not an exempt dwelling, and therefore the first step is to identify the dwelling to be valued.

Section 3 of the 1992 Act sets out the meaning of what is a dwelling:

> 3(2) Subject to the following provisions of this section, a dwelling is any property which –
>
> (a)   by virtue of the definition of a hereditament in section 115(1) of the General Rate Act 1967 would have been a hereditament for the purposes of that Act if that Act remained in force; and
>
> (b)   is not for the time being shown or required to be shown in a local or a central non- domestic rating list in force at that time; and
>
> (c)   is not for the time being exempt from local non-domestic rating for the purposes of Part III of the Local Government Finance Act 1988 ('the 1988 Act').

Section 115 of the General Rate Act 1967 does not offer a clear statutory definition of a hereditament, and simply says that it means

> property which is or may become liable to a rate, being a unit of such property which is, or would fall to be, shown as a separate item in the valuation list.

Further guidance must be obtained from rating case law and the leading case is *Gilbert (VO)* v *S Hickinbottom & Sons Ltd* [1956] 2 QB 40. From this case and others which have followed it, the following principles can be derived:

(1)   The hereditament will be defined by the extent of a person's rateable occupation.

(2)   There must be a single unit of occupation.

(3)   The area of that occupation must be capable of definition- the 'geographic ring fence' test.

(4)   Generally the parts should be contiguous but they do not have to be physically inter connected.

A bed-sitting-room with no facilities other than a wash hand basin, and sharing common bathroom, toilet and kitchen can still be a separate hereditament and therefore be a dwelling for council tax purposes. Sometimes it will be necessary to decide who is in paramount occupation in HMOs (Houses in Multiple Occupation). If it is the landlord because the degree of control over access to the building and restrictions placed on the exclusive rights to occupy accommodation imposed by the terms of the letting agreement, then there will only be one hereditament.

In some cases the hereditament will include non domestic property (business premises) in which case it is called a 'composite hereditament', and the domestic part will be a dwelling for the purposes of council tax by reason of section 3(3) of the 1992 Act. Similar provisions apply in Scotland (section 72(2) of the 1992 Act).

Another definition of a dwelling is to be found in section 3(5) of the 1992 Act, that gives the Secretary of State the power to make an order to provide that anything which would (apart from the order) be one dwelling shall be treated as two or more dwellings (disaggregation); and anything which would (apart from the order) be two or more dwellings shall be treated as one dwelling ('aggregation'). Section 72(5) makes the same powers available in Scotland. In England and Wales these provisions are contained in the Council Tax (Chargeable Dwellings) Order 1992 SI 1992 No 549, as amended in subsequent years .

When there is a single hereditament but more than one self contained unit, the listing officer is required to treat the property as comprising as many dwellings as there are such units included in it. Article 2 defines a 'self contained unit' to mean

> a building or part of a building which has been constructed or adapted for use as separate living accommodation.

The original provisions included caravans and boats, but they were removed with effect from 1 April 1997 so that only buildings or part of buildings can be 'disaggregated' to form separate dwellings. Common examples are 'granny annexes' *Batty (VO)* v *Burfoot* [1995] 2 EGLR 142 and studio flats within a 'foyer' hostel scheme (*Beasley (VO)* v *National Council of YMCAs* [2000] RA 249).

The main test is whether the building or part of a building is constructed or adapted for use as separate living accommodation, rather than the actual or intended use. In the granny annex cases it was noted that there was by the nature of the arrangements a significant element of communal living, but if the annex contained facilities for cooking and preparation of food, eating and sleeping, washing/bathing and a toilet they satisfied the requirements of article 2. In the *YMCA* case Sullivan J acknowledged that it may be relevant to consider the extent of facilities provided in the flat and the remainder of the building, but where all facilities are present for separate living, it may be difficult to argue that the extent of the communal facilities is relevant to the question of whether a room has been constructed for use as separate living accommodation.

Lack of a fixed cooker or bath/shower will not in themselves be fatal to satisfying the requirements for disaggregation but where all that is available in a room is simply the living space, and all other facilities are provided communally, it will not be appropriate to treat the room as a separate dwelling unless it can be shown by the terms of occupation such as letting on an assured shorthold tenancy, that it is in itself a 'single property' (hereditament).

In the *Batty* case it was decided that planning restrictions on the future sale of an annex away from the main dwelling are not matters relevant to the consideration of whether a part of a building is a separate dwelling, but can be taken into account at the valuation stage.

'Separate' can still mean sharing a common hall, stairs and landing (*McColl* v *Subacchi (VO)* [2001] RA 342) but if the only access is through the main living area of another dwelling then the dwellings are not separate (*Batty (VO)* v *Merriman*).

Article 4 of SI 1992 No 549 deals with the question of 'aggregating' more than one hereditament so that there will be a single valuation band. This can only arise when there is more than one hereditament (a 'multiple property') which is occupied as more than one unit of separate living accommodation but consists of only one self contained unit. In this case it is at the discretion of the listing officer whether the units are aggregated, but in exercising his discretion the listing officer shall have regard to the extent, if any, to which the parts of the property separately occupied have been structurally altered (article 4(2)). The greater the alteration to make the various parts more capable of separate occupation then the less likely it is that the listing officer will exercise discretion to aggregate. Terms of the tenancy agreements and fleeting nature of the occupations may also be a relevant factor, and because this provision is similar to the former section 24 of the General Rate Act 1967, which concerned buildings occupied in parts, past cases under the old rating legislation provide further guidance (see *Halliday* v *Melville (VO)* [1964] RA 247).

Valuation assumptions

Section 21 (2) of the 1992 Act says that valuations are to be carried out by reference to a prescribed date of 1 April 1991, and the valuation assumptions were left for prescription later. In Scotland the equivalent provision is section 86(2) of the 1992 Act.

The Council Tax (Situation and Valuation of Dwellings) Regulations 1992 SI 550, subsequently amended in 1994, contains the assumptions for England and Wales. Regulation 6 (2) sets out a list of assumptions, which broadly fall into three groups – financial and legal matters, matters relating to physical circumstances and planning matters.

Regulation 6(1) states that the value of any dwelling shall be the amount which, on the assumptions mentioned in regulation 6(2) and (3), the dwelling might reasonably have been expected to realise if it had been sold in the open market by a willing vendor on 1 April 1991.

'Reasonably have been expected to realise' and 'open market' are not defined in the regulations but it should be assumed that all preliminary arrangements have been made prior to disposal, adequate publicity has been given to the sale, and the most appropriate method of sale has been chosen in order to maximise the sale proceeds.

It is important to determine the value by having regard to all the evidence of sales of similar dwellings in the locality around the period of the valuation date, and then form a view of what might be 'reasonably expected'. If the actual sale price is well above or below that which might reasonably have been expected having regard to the comparable evidence, it should not be followed. To some extent this may be blurred by the need to put a dwelling within a range of value rather than adopt an exact figure, but valuations on the edge of two bands will require greater weighing of the comparative evidence.

Sale prices not tested by the market place are less reliable, so that sales between connected persons such as part of a divorce settlement or as part of a larger financial transaction must be treated with caution. Valuations for the purposes of the 'right to buy' provisions of the Housing Act 1985 are similarly unreliable evidence because they have not been exposed to the market, are often reduced by a large sitting tenant discount, and do not reflect the value of tenants' improvements. If, however, there little or no direct sales evidence then RTB valuation reports before tenants discount are a useful source of secondary evidence provided full details of the report are known.

Most evidence will be obtained from the market in sales of 'second hand' properties, but sales of new dwellings are admissable after adjustment is made to remove any value attributable to incentives such as 'white goods' and chattels, and any new build premium that can be established from evidence of recent resales.

'Willing vendor' is also not defined in the regulations but may be taken to have the same meaning as 'willing seller' in section 5 of the Land Compensation Act 1961 (rule 2). That is to say that the vendor is under no compulsion to sell. The sale is freely negotiated and the property is sold without reserve. Repossession sales by financial institutions following a failure by the owner to maintain mortgage repayments are often regarded with suspicion since the aim of the mortgagor is normally to recover sufficient to redeem the mortgage rather than maximise the return from the sale, and minimise the time taken to realise the asset. If the sale has been freely exposed to the market at auction then the requirement of 'open market' disposal is satisfied but the price may reflect the haste of the building society to secure the resale and hence offend the assumption of 'willing vendor'.

Interestingly, there is no provision in the regulations for a 'willing purchaser'. The definition of open market does not exclude a special purchaser or marriage value, but the extent of any over bid must be established as at 1 April 1991. Special value may also be excluded by assumptions relating to restrictions on development potential.

No reduction in price should be made for any theoretical 'flooding of the market' on 1 April 1991. Regulations 6 and 7 refer to valuation of any one property and do not permit the assumption that all dwellings are placed on the market at the same time. The valuer must consider that the equilibrium of the market place is maintained in the state that actually existed on the fixed valuation date.

The assumptions in regulation 6(2) are:

(a) *That the sale was with vacant possession.*

This means that any sales at a reduced value to reflect the existence of a sitting tenant must be disregarded. In the case of assured shorthold tenancies the discount from full vacant possession value may be quite small, so the evidence will carry some weight if none other is available. Usually the discount for such tenancies is unlikely to exceed 10%. This is in contrast with freehold sales subject to Rent Act tenancies, where the discount is far more significant and less predictable, depending on the prospect to the landlord of gaining possession.

If the dwelling is situated on an estate of similar rented properties then it is permissible to have regard to the actual nature of the occupation of surrounding properties as at 1 April 1991, and any consequential effect on the vacant possession value of the subject property.

(b) That the interest sold was freehold, or in the case of a flat, a lease for 99 years at a nominal rent.

Regulation 6(5) defines a flat as having the same meaning as in Part V of the Housing Act 1985. In essence, a property which forms part of a building will be a flat unless it is divided vertically from the rest of the building and no material part of it lies above another part of the structure.

Sales of short leasehold interests must be disregarded, and valued irrespective of the length of the unexpired term.

(c) That the dwelling was sold 'free from any rent charge or other incumbrance'.

A 'rent charge' is defined in regulation 6(5) as having the same meaning as in the Rent Charges Act 1977, that is an annual or other periodic sum charged on or issuing out of land excluding rent or interest.

'Other incumbrance' will encompass matters of a similar nature or genus, namely a financial restriction which it is within the power of the vendor to remove by his own unilateral action. In the same way that it is possible to redeem a rent charge so mortgages and other financial charges on a dwelling can be redeemed by the vendor, and must be ignored for council tax purposes.

(d) The size, layout and character of the dwelling, and the physical state of the locality, were the same as at the relevant date.

The meaning of 'relevant date' within the meaning of regulation 6(5A) and 6(5B).

If a dwelling existed on 1 April 1993 then it must be valued having regard to its size and character and the physical appearance of the locality at that time, but at levels of value appropriate on 1

April 1991. If the dwelling was built after 1993 the valuer must adopt the size and character of the dwelling when it was built, and transport the locality as it is then back to 1 April 1993, even if in 1993 the locality was just green fields. Similarly, if a motorway is built next to the property it is possible for the valuation band to be reduced to reflect the locality as it now exists, although the dwelling itself may have remained unchanged since 1 April 1993.

If a dwelling is improved or extended after 1 April 1993 and therefore there has been a 'material increase' within the meaning of section 24(10) of the 1992 Act, it is not possible to change the valuation band until it has been subsequently sold as a whole or in part ie a 'relevant transaction' has occurred within the meaning of section 24(10) of the 1992 Act. Section 87(4) (a)(i) contains similar provisions for Scotland. The transfer of a legal share in the property such as a share as part of a divorce settlement or part of a 'staircasing' transaction is not a 'relevant transaction'. If there has been a 'material increase' followed by a 'relevant transaction' then the band can be altered with effect from the date the transaction was completed, and this effective date is the 'relevant date' that determines the physical size, layout and character of the dwelling to be valued.

(e)   That the dwelling was in a reasonable state of repair.

In determining what is meant by 'reasonable repair' regulation 6(6) states that the age and character of the dwelling and its locality must be taken into account. This means that for a house in the middle of a terrace of similar houses its state of repair will be judged against the age and character of the others. If all the other houses have been substantially extended and improved by the addition of central heating, fitted kitchens, new bathrooms, UPVC windows and doors, compared to the one house which is in its original state, then the character of the terrace has changed and the state of repair of the unimproved property must be judged in relation to other similar unimproved houses, not the rest of the terrace.

If a dwelling is in such poor repair that it is incapable of beneficial occupation and being put into repair at reasonable cost, then the entry must be deleted from the list since it is no longer capable of being a 'hereditament' within the meaning of section 3(2) of the 1992 Act.

Disrepair is not a ground for appeal as a 'material reduction' as defined in section 24(10), but lack of repair can be considered when dealing with the compiled list entry, subject to the statutory assumption of reasonable repair.

(f) The common parts in a building such as stairwells and landings in a block of flats are in a like state of repair to the dwellings and the purchaser would be liable to the cost of maintaining them in that state.

(g) Fixtures which are designed to make the dwelling suitable for use by a physically disabled person and which add to the value of the dwelling must be disregarded.

(h) That the use of the dwelling would be permanently restricted to use as a private dwelling.

This is intended to deal with the situation for example, where the majority of former houses in a terrace on the edge of a central business district have been converted to use as offices, and all sales of remaining properties still in residential use reflect the additional value of potential commercial use. In such cases houses which remain domestic property (wholly used as living accommodation) must not be valued by reference to sales evidence derived from commercial use, but by using sales evidence of similar houses in the locality where there is no over bid for commercial use.

(i) That the dwelling had no development value other than for 'permitted development'.

This means that if the open market value of a dwelling with a large garden is enhanced by the prospect of gaining planning consent for the construction of another dwelling on part of the plot, then the development value of that land must be ignored when valuing the existing dwelling for council tax.

'Permitted development' is the small scale development allowed under the provisions of the Town and Country Planning Act 1990, without the need to obtain planning permission.

These valuation assumptions equally apply to composite hereditaments (part domestic and part non domestic properties), with exception of regulation 6(2)(h). Composites are however valued in a different way, and special rules apply (see below).

Valuation method

Capital valuations for council tax are prepared by reference to comparable sales evidence at or around the antecedent valuation date of 1 April 1991, adjusted as necessary to comply with the statutory valuation assumptions set out above. The best evidence will always be that which does not require adjustment, has a sale date close to the valuation date, is situated in the same locality, and is for a dwelling of similar age, size, layout and character to that which is to be valued. Unlike many types of commercial property, sales evidence of dwellings is generally not capable of yielding any meaningful interpretation by seeking to analyse in terms of a common rate per square metre measured internally, although a comparison of gross external areas can be a useful tool of analysis.

New dwellings will be entered in the list from the date they are capable of occupation, or from the date stated in a completion notice, but the value at the date of sale may require to be adjusted if the market has fallen or risen since 1 April 1991. The process of adjustment becomes more difficult as time goes by, so that the sale prices of new houses or resales on an estate built in 2003, which was just green fields in 1991 will have little relevance compared to contemporary evidence at the antecedent valuation date.

In such cases the dwelling must be valued by the nearest comparison with properties that were built before 1991 and for which sales evidence at that time is available. An alternative method is to try and adjust by reference to residential property indices produced by lending institutions. This method is not reliable because monthly or quarterly figures can be distorted by the size of the sample and the mix between different types of dwelling and locations within a wide geographic area. The Valuation Office Agency has a data base of all residential transactions notified for Stamp Duty and Land Registry purposes.

As patterns of value across an area change these cannot be taken into account until a general revaluation unless the change has been caused by some specific change to the physical state of the locality such as the closure and demolition of the premises of a major local employer or the building of new estates that reduce the value of existing dwellings nearby.

The listing officer/assessor will not prepare individual valuations for each dwelling, since it is only necessary to place the dwelling in a valuation band. Because the valuation bands for the more expensive properties are very wide, this will permit wide variations in opinion of the actual value without having any significance for council tax purposes. Only a dwelling whose value falls near the edge of a valuation band, will be sensitive to differences in interpretation of the sales evidence.

There are special rules that apply to composite hereditaments, which are contained in regulation 7 of SI 1992 No 550. In these cases the value of the domestic part shall be taken to be that portion of the 'relevant amount' which can reasonably be attributed to domestic use of the dwelling.

'Relevant amount' means:

> the amount which the composite hereditament might reasonably have been expected to realise on the assumptions mentioned in regulation 6, other than paragraph (2)(h) of that regulation...

The first step having defined the hereditament is to assume a sale of the whole composite hereditament as one unit of occupation, before apportioning the value 'reasonably attributable to the domestic use of the dwelling' in order to arrive at the value of the dwelling. Because regulation 7(1) requires a band to be ascribed which reflects the value which can be reasonably attributable to domestic use of the dwelling the distribution between domestic and non domestic use of a composite hereditament should reflect how the market would view its use if it were made available with vacant possession, not necessarily the use made of the property by the actual occupier.

Where a clear pattern of use is discernable, such as the use of upper parts over a parade of shops, this should be followed, but any distribution between domestic and non domestic uses which does not conform to the prevailing pattern of use in that locality should be disregarded. In practice, it will be appropriate in most cases to have regard to the actual use as opposed to any notional use.

There is no requirement to carry out an actual valuation of the whole in order to determine the 'relevant amount'. The listing officer must have regard to the likely value of the whole, within a range of value, but having analysed the sales evidence for dwellings which have been sold separately compared to those which have been sold as part of a larger composite hereditament, he is entitled to arrive at a scheme of valuation based on the 'separately lotted value' of the dwelling less a deduction for being part of a composite hereditament, the size of the discount being determined by the evidence.

This approach was approved by the Court of Appeal in *Atkinson and others* v *Lord (VO)* [1997] RA 413, which concerned the valuation of farmhouses in Cumbria.

Special valuation cases

Public houses

The valuation of those parts of licensed premises used for domestic purposes will have regard to characteristics of the public house itself. In order that the band ascribed to the domestic use fully reflects that it is a portion of the value of a larger hereditament which itself is valued by reference to its trading potential, it is necessary to consider the actual/notional accommodation, trade and location.

The living accommodation in a public house will always have a value to a prospective purchaser, either as somewhere to live or in order to provide accommodation for a manager. Irrespective of the level of trade of the public house, this value will not fall below the lowest value which attaches to similar alternative accommodation in the locality which could be used to fulfil this purpose This level of value is called the 'minimum band'. An example would be the domestic use of upper parts of a public house situated in a city centre location. Values of similar size flats which would be acceptable to prospective purchasers ranged from £50,000 to £60,000 at 1 April 1991. If an adjustment of 10% is made for the disadvantages of being part of a pub eg noise and disturbance or no separate access, then the value of the domestic use of the public house would not fall below £50,000 − 10% = £45,000, or band B in England. This is known as the 'minimum band'.

The 'maximum band' will be band C using the same example – value of the living accommodation would not exceed £60,000 − 10% = £ 54,000, where the value of flats of a similar size and quality in the immediate vicinity is £60,000 as at 1 April 1991.

In locations where location determined that there was a difference between the minimum and maximum bands it is necessary to consider the trading potential of the public house and determine a 'trading band'. This is found by having regard to the capital value of the public house at 1 April 1991. After an analysis of market evidence and 1990 list rateable values of public houses that were sold during the year preceding and following the antecedent valuation date, a table was compiled by the valuation office as shown below:

1990 rateable value	Likely capital value range	'Trading band'
Up to £6,000	Up to £200,000	A
£6,001–£9,000	£150,000–£300,000	B
£9,001–£12,500	£250,000–£450,000	C
£12,501–£17,500	£400,000–£600,000	D
£17,501–£25,000	£500,000–£800,000	E
Over £25,000	Over £750,000	F

In cases where the minimum and maximum bands are the same the band letter will represent the value attributable to domestic use.

Where the minimum and maximum bands differ, the value to be ascribed will never fall below the minimum band. Where the trade band is higher this will be adopted, subject to not exceeding the maximum band unless there are circumstances such that the purchase of the public house is primarily motivated by its domestic use and situation as opposed to the business as a whole.

Agricultural dwellings

The first step is to decide whether the dwelling is part of a larger composite hereditament (the farm) or whether it is a separate hereditament in its own right. The majority of farms will comprise of land, buildings and a farmhouse all of which are occupied by one person, ie the farmer. In such cases all the land and buildings which fall within a geographic ring fence will form the hereditament, and it will be necessary to apportion the 'relevant amount' in order to arrive at the value of the farm house.

The farm may also include agricultural workers cottages, and it is necessary then to decide whether they form part of the main holding in the paramount occupation and control of the farmer, or are occupied quite separately by a farm worker or by a person with no connection to the farm.

If the cottage has been sold off or let to someone not connected to the farm, then the cottage will be a separate hereditament and valued just like any other dwelling, reflecting any advantages or disadvantages of its location and proximity to farm buildings. If the cottage is occupied by a farm worker, and it is part of the contract of employment eg a cottage occupied by a herdsman who is obliged by his contract of employment to live there to be in close proximity to the herd at all times, then the farmer is in 'rateable occupation' and the cottage will form part of a composite hereditament. It must therefore be valued as part of a composite hereditament.

Houses and cottages situated away from the main holding will generally not be part of the hereditament unless it can be shown that the cottage is functionally essential to the farm. This is unlikely where the cottage is located in the centre of a village rather than close to the farm. It may however, be necessary to discount the value to reflect a planning permission which restricts the use and occupation to a person engaged in agricultural employment in the locality, in which case a reduction in order of 10%–30% can be made from the sale price without such a restriction, depending on local sales evidence in 1991 regarding the effect of a restrictive planning condition on value.

It is not appropriate to make an allowance for both a planning restriction and being part of a composite hereditament.

Valuation bands

The bands to be ascribed to all dwellings in accordance with sections 5(2), 5(3) and 74 of the 1992 Act vary between England, Wales and Scotland. The tables below set out the bands for each.

In England

Valuation band	Range of values
A	Not exceeding £40,000
B	Exceeding £40,000 but not exceeding £52000
C	Exceeding £52,000 but not exceeding £68,000
D	Exceeding £68,000 but not exceeding £88,000
E	Exceeding £88,000 but not exceeding £120,000
F	Exceeding £120,000 but not exceeding £160,000
G	Exceeding £160,000 but not exceeding £320,000
H	Exceeding £320,000

In Wales

Valuation band	Range of values
A	Not exceeding £30,000
B	Exceeding £30,000 but not exceeding £39,000

C	Exceeding £39,000 but not exceeding £51,000
D	Exceeding £51,000 but not exceeding £66,000
E	Exceeding £66,000 but not exceeding £90,000
F	Exceeding £90,000 but not exceeding £120,000
G	Exceeding £120,000 but not exceeding £240,000
H	Exceeding £240,000

In Scotland

Valuation band	Range of values
A	Values not exceeding £27,000
B	Values exceeding £27,000 but not exceeding £35,000
C	Values exceeding £35,000 but not exceeding £45,000
D	Values exceeding £45,000 but not exceeding £58,000
E	Values exceeding £58,000 but not exceeding £80,000
F	Values exceeding £80,000 but not exceeding £106,000
G	Values exceeding £106,000 but not exceeding £212,000
H	Values exceeding £212,000

Chapter 8

Cables and Telecommunications
Installations

The Telecommunications Act 1984 (the 1984 Act), led to substantial growth in the number of radio mast sites and telecommunication cables across the country. Until then radio communication was confined to such operators as BT, and non-telecommunication operators such as the MoD, local authorities, etc.

From the mid 1980s the rapid growth and use of radio tele-communications, especially mobile phones (which had over 40 million subscribers in the UK by 2001) led to a massive expansion in the number of radio mast sites. Radio masts now range from multiple antennae on masts and high buildings (macro cells) to small unobtrusive fixtures on the faces of, or inside, buildings (micro cells). It is predicted there may be as many as 80,000 tele-communication base stations (three to four times as at present) by about 2010. The greatest concentrations will be in urban areas and along main transport corridors. Coupled with this expansion and the technological revolution, there has also been a corresponding growth in telecommunication cable networks across the country.

The market is relatively specialised and, while still evolving, trends and unique valuation issues can be identified. Most telecom-munication sites are leased but a capital market exists and is dealt with separately. The evidence of telecommunication mast rentals has provided a basis for valuing non-telecommunication masts.

There are two distinct areas of the market:

* Radio masts.
* Cables and wires.

Both are subject to the same legislation but have, historically, evolved differently.

Telecommunications Act 1984

The 1984 Act requires telecommunications operators to be licensed by the Department of Trade and Industry. To facilitate the

143

installation of their system, certain operators are granted 'Code Powers', the terms of which are set out in Schedule 2.[1] If exercised the Code confirms on operators the right to install and maintain apparatus on private land. For the purposes of public highways the so-called 'Code System Operators' are 'undertakers' for the purposes of the New Roads and Streetworks Act 1991.

Paragraph 2 of the Code states that the agreement of the occupier has to be obtained in writing to execute any works on land for telecommunications purposes, to keep apparatus installed and to enter the land to inspect the same. The agreement can only be exercised in accordance with the terms in which it is conferred. It is particularly important, therefore, that any agreement adequately sets out the rights and obligations of the parties to avoid prejudicing either and to regulate the future conduct of the relationship.

Paragraph 5 contains compulsory powers to acquire rights over land. The operator must give notice of the need for such rights and, after 28 days, may apply to the court[2] for an order to dispense with the need for landowners consent. The court must grant an order where it is satisfied that the effect of the order can be properly compensated for by money and that the loss to the grantor is outweighed by the benefit to others.

Paragraph 7 deals with the financial terms. This provides that there must be:

- consideration in respect of the giving of the wayleave as would have been fair and reasonable if the rights had been given willingly; and
- compensation for any loss or damage suffered by the grantor (including injurious affection).

The word 'consideration' is a fundamental departure from previous compulsory purchase legislation, which has tended to be on the basis of compensation (ie profits or losses foregone).

After overhead apparatus (either new or substantially different from a previous installation) is installed, an owner or occupier either of the land on which it stands or of land which is nearby and where the enjoyment of that property is substantially affected can make formal objections (under paragraph 17) seeking relocation. If

---

[1]   From 25 July 2003 it is anticipated that this will be known as the 'Electronic Communications Code' and will be applicable to all relevant persons (on application) under the Communications Act.

[2]   County Court in England and Wales. Sheriff Court in Scotland.

pursued formally, a court can order changes but it is believed that this would only apply in exceptional circumstances.

Paragraph 21 of the Code restricts the landowner's ability to require removal of any telecommunications equipment installed on his property. A Code System Operator can serve a counter-notice within 28 days of a notice to remove and, in such circumstances, the landowner can only effect removal with the consent of the court. This provision is particularly pertinent where landowners may be prejudiced by the continued existence of telecommunications equipment after the agreed term and should be pointed out to clients. Provisions regarding compensation set out in the Code may protect the landowner's interest in circumstances such as the future development of the property.

Where apparatus has been abandoned or is not likely to be used again, the operator is obliged to remove the telecommunications apparatus under paragraph 22. The potential costs of decommissioning sites should not be underestimated. It is open to the owner to take over the equipment and to agree a payment in lieu of its removal.

The first time the legislation appears to have been considered by the courts was *Mercury Communications Ltd v London & India Dock Investments Ltd* [1994] 1 EGLR 229. This case related to a fibre optic cable serving Canary Wharf. In it, the judge, Hague QC appeared to draw a distinction between open market and 'consideration', stating:

> ... What I have to determine is not the same as what the result in the open market would have been if the grant had been given willingly. That is, however, far from saying that the market result is irrelevant or can afford no guidance. Indeed, in my view, the market result is the obvious starting point; and in most cases it will come to the same thing as what is 'fair and reasonable', because *prima facie* it would be neither fair nor reasonable for the grantor to receive less than he would in the market or for the grantee to have to pay more than he would in the market. But there may be circumstances, of which the absence of any real market may be one, in which a judge could properly conclude that what the evidence may point to as being the likely market result is not a result which is 'fair and reasonable'...

The case appears to hold that compulsory purchase principles, including the *Pointe Gourde* principle[2] are not applicable. A ransom

---

[3]   *Pointe Gourde Quarrying & Transport Co Ltd* v *Superintendent of Crown Lands* [1947] AC 565 PC, 63 TLR 486.

value approach, having regard to the profit which the operator was forecasting for the use of the cables, was also considered inappropriate. The evidence of terms of aerial agreements with telecommunication operators was considered to be of general assistance in showing the bargaining power of grantors and grantees for cable leases. It was stated that agreements with the National Farmers Union (NFU) and Country Land and Business Association (CLA) were of no value in this case because they related to agricultural land and did not include anything for the bargaining power of the grantor. The court considered that the best evidence for the case related to the laying of cables.

It would appear from the arguments put forward in this case that reference to 'ransom' was in regard to the profitability of the line and not to the bargaining strengths of the parties as was the case in *Stokes* v *Cambridge* (1961) 13 P&CR 77.

Mercury was a county court judgment but the valuation principles it set out were considered by the Court of Appeal in *CableTel*[4]. They observed that the exercise required under paragraph 7 of the Code is not necessarily one of ascertaining market value although this is a relevant consideration. What was relevant in fixing 'fair and reasonable value' were the advantages which an operator could obtain from the rights sought. The Court is entitled to take into account the potential commercial profitability for an operator as a factor.

Radio masts

It is sometimes argued that the radio mast market is relatively unsophisticated. Often acquisition agents are unqualified and work within guidelines dictated by operators. There is however clear evidence in radio mast negotiations of different rents being offered to achieve differing lease terms. In any event it is a surveyor's job to make subjective valuation judgments based on market evidence.

There is no evidence in the market place that the contractor's method, deriving value from the cost of the installation, is appropriate. Sites with quite disproportionate build costs often command the same rent in the market. Rental assessments are normally based on comparable evidence following commercial rental practice.

---

[4]　　*CableTel Surrey and Hampshire Ltd* v *Brookwood Cemetery* [2002] EWCA Civ 720.

The majority of agreements are agreed in situations where the landlord often has very little knowledge of this specialist market and has often not been advised. It is the practice in the telecommunications market that standard terms are offered by the tenant and that the tenant's solicitor prepares leases. Such terms can be considerably improved on negotiation. Clients should be made aware that it is normal for operators to pay some, if not all, the landowner's reasonable legal and surveyor's fees in respect of such negotiations.

As well as telecommunication operators, there are a number of companies operating radio masts, etc. who fall outwith the 1984 Act. Historically the rents for such sites were based on compensation principles rather than market value because of the lack of market evidence. The market value differential between these sites and those leased to telecommunication operators has narrowed over the years as the growth of telecommunication sites has helped establish a solid base of evidence:

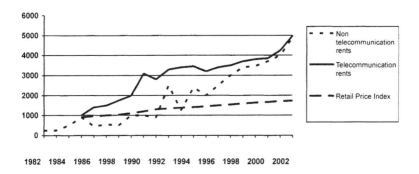

Source: Strutt & Parker Research

There are three distinct areas in the rental market for radio mast sites:

- Greenfield sites.
- Rooftop sites.
- Site sharing.

Greenfield sites

Telecommunication masts tend to be erected along major road and rail networks in the form of cells, some 6–10 km apart. Non-telecommunication masts are often located on hilltops to provide wider coverage.

Most sites have the same basic equipment. An aerial, usually a steel pole or lattice, supports antennae and/or dishes. A cabin or cabinets contains ancillary equipment. The dishes (or a telecommunication cable) connect the site to other sites and/or to the fixed network.

The principal factors that affect the valuation are as follows.

Competition

When any mast operator offers a rent for a particular site in the open market, there is rarely another operator seeking a mast in the same location because of differing network requirements. 'Competition' for a particular site is not necessarily a factor. Attention is drawn to *Mercury Communications Ltd* v *London & India Dock Investments Ltd* [1994] 1 EGLR 229A–F in relation to this.

Location

Radio masts tend to be part of a network and there are licence obligations on telecommunications operators to provide national coverage.

There is little rental evidence to support the case that low throughput sites in remote rural areas command any less rent than those in more populous locations. Premium rentals may however be paid for very high throughput sites such as within the M25, near built-up areas, and within high population densities etc.

Planning consent

Telecommunication masts under 15 metres originally fell under permitted development rights and did not require planning consent. This remains the case in England, but not in Scotland where most new masts require planning consent.

Planners are reluctant to permit the proliferation of new masts and it can be argued that sites with existing consents have increased in value. A mast with an extant planning consent can be extended to a limited extent through permitted development rights.

Generally the taller the mast the greater the value; perhaps because the more difficult it is to obtain planning consent. This is clearly illustrated by a recent survey, but is less pronounced than it was five years ago.

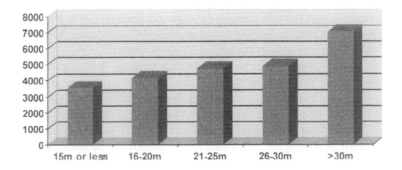

Source: Strutt & Parker Research

Above about 50 metres, most masts tend to be 'live'. That is to say the mast itself is the antenna, rather than a support, and, in such instances, the height of the mast is governed by broadcasting frequency and there is little correlation between height and rent.

Area

There is little direct correlation between the area of a radio mast lease and rent. In quantum terms, given that the alternative use value is most often agricultural, the size of compound tends not to be a major issue with landowners but may be the case where space is at a premium such as in urban locations. Operators are usually willing to adjust design to accommodate smaller compounds where space is at a premium.

Equipment restrictions

It is common for leases to include restrictions on the overall height and on the number of antennae and dishes permitted. This seems to be due to the inter-relationship of rent and height and the visual effect of the installation on neighbouring property

Operators usually seek wide rights to alter the apparatus as business grows and there is evidence to suggest they will pay a premium for such leases.

Rent review provisions

Operators tend to seek five-year review patterns. Strutt & Parker's Telecommunications Survey suggests that 70% of sites are however reviewed three yearly, with only 19% five yearly.

Operators tend to seek reviews in line with the RPI. Radio mast rentals have, however, outstripped the RPI by a significant factor in recent years, but as operators get closer to achieving full coverage the number of new sites required is likely to fall. It is probably for this reason that most agent represented deals are based on the higher of open market rent or RPI and are usually upwards only:

Source: Strutt & Parker Research

Alienation

Assignation is normally prohibited outside the company group. Most operators now insist on a right to assign subject to landlord's consent (such consent not to be unreasonably withheld or delayed). This may mean that a landlord is unable to obtain a premium for the granting of consent[5]. With payments being made to landlords in

---

[5]   Landlord and Tenant Act 1927 section 9 in England. In Scotland note *Renfrew DC v AB Leisure (Renfrew) Ltd (in liquidation)* SCT 635.

respect of consent to assign currently in the order of £3,000–5,000/mast, this is a valuable right that falls to be considered as part of the rent.

Site sharing is the practice whereby third parties install and operate their antennae and/or dishes on existing masts. As planning authorities become more concerned about the proliferation of radio masts across the country, they may try to insist that new operators seek space on existing masts. It is a condition of most licences granted under the 1984 Act that operators seek means other than erecting new masts to fulfil their site requirements and that they permit site sharing. National planning policy guidelines and Codes of Practice reiterate this[6]. The potential income to the mast operator from site sharing can be considerable and it is normal for landlords to restrict site sharing. Landlords hold the 'golden key' to this income: see *Stokes* v *Cambridge supra.*

It is fairly standard for new leases for telecommunications purposes to comprise a rent plus an additional stated percentage of any site sharing income taking place. Up until about 1996 the normal 'payaway' to a landlord was 50% of the site sharing income received by the head tenant, but operators have been successful in driving down this percentage and it is now usually not less than 30%. In new masts, if satisfactory terms cannot be agreed in respect of 'payaway' for permitting this, site sharing is usually prohibited. Consideration should be given as to whether or not the payaway is net of deductions, such as management charges, etc, or gross. Parties to such an arrangement should be aware of situations where site share rental is reduced because of other considerations (such as capital contributions to mast replacement or reciprocal arrangements).

Where there is no restriction on alienation the potential for the tenants to generate additional income from site sharing over and above their own use of the site falls to be considered as part of the base rent. The valuer should consider the likelihood of site sharing income in the period to the next review. Allowance for this potential should be made with appropriate adjustments for the length of term and whether or not the review clause is upwards only.

Valuers should be aware of the various facets of site sharing covered below when advising the landlord in respect of site sharing on radio mast sites.

---

[6]   The planning system however makes this a difficult area since decisions must relate to the subject site and be decided on individual merit.

Termination

It is normal to grant the tenant a right to break within 24 months of a new lease since site locations are computer generated. A penalty clause for early termination is often included.

Some radio mast agreements permit tenants wide termination rights. Most lock the operator into a fixed lease for 10 to 20 years. Operators want security for a minimum period of about five years to recoup their investment.

Landlords are reluctant to accept break clauses other than on the 10th anniversary whereas operators seek flexibility. Given the rapid technological developments in this industry, a favourable termination clause is of considerable value to a tenant and operators have offered up to 25% of the passing rent to obtain such breaks.

Many urban telecommunication leases include the ability for the landlord to beak in the event of redevelopment. The effectiveness of these should be considered carefully in the light of Code powers.

Premiums

Often premiums are offered to sweeten deals and/or to assist in speedy legal completion. These are often called 'early access' or 'disturbance' payments and may be set out in documentation separate from the main lease.

There is also evidence of premiums also being paid as part of rent review settlements.

Valuers are reminded of RICS guidance notes in respect of such payments.

Rooftop sites

The density of telecommunication sites required is far greater in urban than in rural areas and telecommunication sites may be as little as one-third of a mile apart. Most sites are situated on rooftops but smaller microcells covering limited areas are common. About 67% of all telecoms base stations are currently located on buildings and this is set to intensify further with the advent of the latest '3G' technology.

Rooftop rentals are more akin to site sharing rentals than greenfield sites, possibly because several operators can be accommodated on one roof and support for the antennae is provided by the landlord.

When masts or equipment are attached to or placed on a building, specific Building Regulations requirements for any alterations to the building may have to be met (eg if the roof structure has to be strengthened or access is provided).

Limited permitted development rights exist, particularly for micro cells.

The market can be divided into:

Micro cells

These tend to a small box the size of a burglar alarm connected to a cabinet providing coverage for a radius of a few hundred metres. Micro cells can be incorporated into street furniture, such as lamp posts, etc. As stated rights are granted within the 1984 Act for Code System operators to install equipment in the public highway. Presently this can be done free of charge.

Rents are currently in the order of £2,500. As in rural sites, operators seek five-year RPI review provisions, but significant improvements can be negotiated. Because the choice of site does not tend to be restricted, values are not as high as for macro cells.

Even smaller are the picocells installed within shopping malls, etc. These are the size of smoke detectors and have limited capacity. A network of such cells within buildings are linked to a central node. Rents for such networks are in the order of £4,000–5,000 pa depending on the number and type of picocells involved.

Macro cells

Macro sites provide area coverage. The choice of site tends to be more restrictive.

Antennae are fixed either on to a small stub tower or mounted directly on to the building. Antennae can be hidden behind fibreglass shrouds blended to match the structure, such as behind the louvres in church towers or as false chimneys, etc. This method is adopted particularly in conservation areas where there are restrictive planning policies.

Equipment cabins are either installed on the roof or at ground level or the equipment housed in surplus space within the building. Given the weights involved care should be taken to ensure that the structure is not compromised.

Factors affecting rents are similar to greenfield sites. The main factor is the location and the number of competing sites. The best/

tallest building in a particular area may attract premium rents.

Rents may be based on a per dish/antennae basis. Generally the rent per unit in such leases is higher as it avoids payment for unused potential.

Size of the equipment is a factor (as in site sharing).

Site sharing

Radio masts can be shared to different degrees. The usual method of site sharing is the use of the radio mast structure to support equipment belonging to third parties. Antennae and dishes are fixed to the mast and ancillary equipment may be installed in existing buildings or the site sharer may have their own cabin within the site. Another level of site sharing is the sharing of single antennae or co-operative agreements between operators (roaming agreements).

Site sharing agreements are usually constructed as licences for periods up to 10–15 years and normally specify the quantity and type of equipment and its location on the mast. Telecommunication companies have national agreements between themselves, which they tend to treat as confidential. The concept of site sharing has been derived from English practice and may be held to be subtenancies by a Scottish court, given differences in the concept of a licence: see *Brador Properties* v *British Telecommunications plc* 1992 SLT 490.

Site sharing rentals tend to be based on standard rate cards according to the type and height of equipment installed, often derived from the wind load effect on the structure. The methodology and unit rate vary between operators but currently rents are in the order of £1,500–2,200 per unit. A unit generally equates to a single antennae. The larger the equipment or the higher up the mast, the greater the number of units and the correspondingly higher the rent. Rents tend to be reviewed annually or triennially in line with the RPI for the period of the licence. Examples of a rate card are found at Appendix A.

Telecommunication operators appear to charge lower rates between themselves than do third party operators perhaps because of reciprocal agreements. Operators may offer discounts in respect of bulk (either by the number of sites and/or on the amount of equipment on one site). Where the site sharer contributes significant capital to enable site sharing to take place (eg by replacing a mast) there is often a rent free period. These issues should be borne in mind when valuing site sharing rights.

Some mast operators also attempt to recharge any payaway to a landlord back to the site sharer. This practice should be considered carefully in the light of the terms agreed in the head lease.

Capital market

There is interest from speculators and investors in purchasing radio masts. Some site operators may be keen to purchase in order to capitalise on site share potential.

The sitting tenant is likely to be a potential buyer where the lease terms are unfavourable to them and there is potential for substantial rent increases. They are however constrained by internal financial constraints instigated by funding institutions. Investors seek sites where there is potential for income growth.

Radio masts are but one form of investment competing for funds and therefore performance must also be measured against other investments in terms of current yield, rental and capital growth. The problem in comparing the subjects with gilts or debentures is that few go beyond about 30 years. Radio masts compete with other property deals as an investment and compare very favourably with yields and income obtainable from other forms of rural property which is probably why so few radio masts come onto the capital market.

Rent increases, site sharing and the potential for site upgrades are three areas that can give rise to significant growth in the income obtained by an investor from a radio mast site. Beyond these instances there are clearly opportunities at the end of the lease and from potential deals on reinstatement provisions when a tenant decommissions a site.

Values tend to be based on normal cash flow appraisals. As a broad rule of thumb multipliers of 13 times rent are frequently adopted but when significant site sharing income is involved this may fall (because of a perceived risk of site sharers leaving).

Cables and wires

Cables and wires used for public telecommunications purposes also fall within the 1984 Act. The valuation of these appears to have been distorted by market practice over the years but is none the less evolving.

Historically the main operator has been BT, who have a well-established network of lines both above and below ground. Fibre optic networks joining main cities and towns in a series of loops developed in the mid 1990s. In these data is transmitted using glass fibres rather than electrical circuits. Fibre is able to carry more data than metal cable, is thinner and lighter and transmits digitally (the natural form for computer data) rather than in analogue format.

BT entered into a standard agreement for rent and compensation with the landowning and farming bodies [the CLA for England, the Scottish Landowners Federation (SLF) for Scotland and the NFU]. This practice mirrored agreements for compensation arising from poles and underground lines in the electricity industry. Other Code operators also agreed rates (eg the Electricity Supply Industry (ESI) for fibre optic cables installed on pylons). It should be noted that electricity companies often erect fibre, etc, on their apparatus as part of their electricity undertaking but as soon as this equipment is used for telecommunication purposes (as defined by the 1984 Act) a further wayleave with the landowner is required. Some electricity companies are also promoting their structures for the installation of telecommunication antennae. These rights should be valued in the light of site sharing rents as set out above.

Telecommunications rights however involve not just 'compensation' but also 'consideration'. Judge Hague QC, in *Mercury Communications* commented that the agreements reached with the CLA/SLF/NFU tended to be based on the agricultural value and made no allowance for the bargaining strength of the parties. Both *Mercury* and *CableTel* emphasised the importance of the benefits the operator expected to gain from the wayleave as being a significant valuation factor. The evidence led in *CableTel* made a clear valuation distinction between local loops and trunk networks which was approved at appeal.

The CLA/SLF/NFU have always stated that such rates are not binding on landowners, but in practice the operators will not make any payment that they deem to be outwith these rate cards. For instance BT categorically states that such rates are not negotiable. This has tended to 'fix' payments unlike radio masts where market forces have evolved.

In instances where landowners have been free to bargain, rates of between £1.00 and £3.50 per metre pa have been obtained for underground fibre optic cables in rural areas.

Given the entitlement in the Code to compensation as well as consideration, it is arguable whether there should be any differential between overground and underground cabling.

Much of the fibre optic cable network is laid within the public highway under powers granted to licensed operators in the 1984

Act. Careful consideration should be given to the extent of the highway in each instance. The local highway authorities currently can seek no consideration from operators for the installation of equipment in the highway. The threat to lay cable in this manner tends therefore to curtail the consideration paid to private landowners for the right to lay cable.

Fibre optic cables also require repeater stations along the line. These are similar to the cabins associated with radio masts and appear to command similar rents.

Operators often seek to capitalise annual agreements based on a multiplier of any annual payment. BT currently offer multipliers of 15 times annual payment as standard. Any likely claim for injurious affection and the prospect for rent increases should be considered as this is not usually reflected in such offers.

Some operators offer rates based on capitalised payments agreed with the CLA/SLF/NFU. Some of these are summarised at Appendix C.

In both the lead cases the courts awarded a once-and-for-all payment rather than an annualised payment. In *CableTel*, the Appeal Court found that the judge was right to order a once-and-for-all payment. The bulk of telecommunication wayleaves are however based on annual payments.

Permanent agreement to retain apparatus would probably not prevent a subsequent application being made to alter the apparatus under paragraph 20 of the Code, but deprives the owner of any other right to end the arrangement.

Frequently agreements are documented in standard formats provided by the operators. These are sparsely worded. Often the yearly payments and the rights granted are not adequately defined. Frequently there is no definite mechanism included for the termination of the agreement or provision for the initial rate to be altered. Given the nature of the agreement terms should be properly and fully documented.

© Ian Thornton Kemsley

Example Site Share Ratecard

Most operators have standard rate cards. An example of one such ratecard illustrating how site sharing rental is derived is set out below.

The number of units tends to be a function of the type of equipment and the height on the mast:

			Units					
						Dish		
		Type of Antennae				(diameter in metres)		
Height on mast	Dipole	Collinear or Yagi	Sector or Panel	0.3	0.6	0.9	1.2	1.8
0–20 m	0.6	0.7	0.8	0.6	0.8	1.2	1.6	3.0
20–35 m	1.0	1.1	1.2	1.0	1.2	1.6	2.0	3.4
36–45 m	1.4	1.5	1.6	1.4	1.6	2.0	2.4	3.8
45–60 m	1.8	1.9	2.0	1.6	2.0	2.4	2.8	4.2

Allowance is made for cabins/cabinets:
Cabins >12m^2 0.4 units
Cabinets minimum   0.2 units

From this a rent is calculated using a standard rent per unit

Example:

3 panel antennae @ 15 m =   2.4 units

    Site share rental 2.4 units @ say £1,800/unit = £4,320 pa

1 Collinear antennae @ 31 m =   1.1 units
2 Colinear antennae @ 25 m =   2.2 units
0.9 m^2 cabinets =   0.2 units

    Site share rental 3.5 units @ say £1,800/unit = £6,300 pa

Annual Agreements with CLA/SLF/NFU

BT

BT (like any other telecommunications operator) has a duty to secure the agreement of the occupier but a landowner is not bound by that consent unless the lease is for a period of more than seven years. BT usually negotiates with the owner to be sure of their position. When the owner is also the occupier, a single annual payment is made. Where there is a tenant in occupation, BT make a full annual payment to both landlord and tenant.

BT make no payment in respect of lines crossing land in the same ownership as the dwelling serviced by the equipment. Their argument appears to be based on the additional value to the owner their service affords. This approach is under scrutiny.

BT has an obligation under statute to provide services. Unlike other operators BT are unable to walk away from unprofitable operations or proposals.

Equipment	Rental (£ per annum)			Compensation (£ per annum)			
	2003	2004	2005	Arable	Permanent Pasture	Rough	Hedges
Single pole	£7.56	£8.16	£8.82	£10.60	£2.40	£1.10	£1.00
Pole and Strut /pole and stay	£8.19	£8.84	£9.55	£17.50	£4.35	£2.00	£2.00
Stay	£7.56	£8.16	£8.82	£13.25	£3.80	£1.20	£2.10
Duct and associated cable	£0.65 per metre	£0.65 per metre	£0.70 per metre	–	–	–	–
Joint box	£26.40	£26.40	£26.40	–	–	–	–

These rates are recommended to their members by the SLF/CLA/NFU. They are not binding and special rates should be negotiated to reflect particular circumstances.

ESI

The CLA/NFU reached agreement with the Electricity Supply Industry for fibre optics on pylons in 1999 for a period up to November 2004. The SLF/NFU in Scotland entered into a similar deal the following year incorporating one extra year.

The negotiations are based on the existence of the electricity apparatus to support the fibre and that the operator owned the pylons on which the cables are strung. Frequently however the fibre optic operator is a third party and the standard wayleave sought contains much wider rights. Such wayleaves should be carefully scrutinised.

Rates were agreed based on pylon size. Standard rates per metre basis are:

	2002–2003	Rental (£ per metre) 2003–2004	2004–2005 (Scotland only)
Up to 36 fibres	£0.315	£0.315	£0.345
Unlimited fibres	£0.450	£0.450	£0.495

These rates are recommended to their members by the SLF/CLA/NFU. They are not binding and special rates should be negotiated to reflect particular circumstances (such as when a fibre optic cable serves a radio mast).

Appendix C

Capitalised Agreements reached by operators with CLA/SLF/NFU

Mercury

The last CLA/NFU agreement with Mercury was based on a capital payment for the installation of 4 ducts in single trench for a 20-year term. This agreement ended in 1999 since when it has not been reviewed.

| 20-year easement | Payment | Disturbance | |
| | | Arable | Pasture |
Owner			
4 ducts	£6.00 per metre	N/A	N/A
Additional duct	£1.00 per metre	N/A	N/A
Chamber not			
exceeding 1 × 2 m	£226.50	N/A	N/A
Occupier			
4 ducts	N/A	Nil	Nil
Additional duct	N/A	Nil	Nil
Chamber not			
exceeding 1 × 2 m	N/A	£355	£85.80

The whole was subject to a minimum payment of £300. In addition to the above compensation was paid for any crop loss etc caused by the execution of the works.

The current CLA/NFU advice is to base an agreement on that reached with 186 k below.

186 k

The CLA/NFU reached agreement with Transco for a fibre optic network to run along existing gas wayleaves in England in September 2000. The line was later vested in 186K and the network was subsequently sold to Hutchinson Network Services.

The SLF/NFU in Scotland refused to endorse the deal (but the rates were applied by 186k none the less).

The CLA agreement was based on draft documentation containing a consent form, deed of easement and code of practice. The CLA stressed that operators would seek to bypass if objection was raised to the package given the speed at which 186k intended to lay the line.

	25 year easement with option to renew for 25 years		Permanent Easement
Owner	First 25 years	Option to extend	
4 ducts	£9.00 per metre	Same price + 3%	£12.00 per metre
Additional ducts	£2.00 per metre	compound pa	£2.00 per metre
Junction box	£400 each	over 25 years	£400 each
Occupier			
365 days		£2.80 per metre	£2.80 per metre
or more		(min £250)	(min £250)
Less than			
365 days		£250	£250

In a number of instances further consideration was agreed by means of side agreements.

Chapter 9

Plant and Machinery as a Loan Security: A Case Study

Introduction

To many professionals involved in the property industry, plant and machinery valuers remain practitioners of a black art, not least because they are often perceived as oily, bearded, beer-swilling monsters. The reality is that they are rarely oily, seldom monstrous, and most shave occasionally. In professional practice, the level of expertise needed is at least equal to that required of those employed in the valuation of real estate. The subject-matter necessitates modification of the techniques employed, but in essence plant and machinery valuation requires the same processes of analytical comparison. However, although the discipline is viewed as a niche, the sphere in which the plant and machinery valuer operates is a wide one, since it encompasses every item of equipment used in any industry. In addition, there is also a large range of purposes for which a valuation might be required.

The purpose of this chapter is to illustrate some of the techniques employed by plant and machinery valuers, hopefully to strip away the mystique and replace it with respect for the professionalism employed. Fellow practitioners will probably gain little from the following pages: most will have been engaged on higher profile, more difficult, more interesting jobs. The writer does not claim any expertise or professionalism above that which could be expected from any property professional engaged in earning a living through his skill and knowledge in valuation. This chapter is intended for non-practitioners, to give them an insight into the everyday work of a typical plant and machinery valuation. It is meant to achieve this by way of a case study of a valuation for one particular purpose that will illustrate the implementation of typical valuation practices. In the following pages, paragraphs in italics are extracts from a plant and machinery report. Normal text is an explanation of the investigative and analytical processes that are summarised in the report.

Case study

Instructions

In accordance with your instructions, we have attended on site in order to carry out a valuation of the plant and machinery assets on the basis of Open Market Value of individual items for removal from the premises at the expense of the purchaser. We understand the valuation is required to assist you in the consideration of loan facilities to be secured upon this plant and machinery.

In any valuation, it is important to establish what the client requires from the beginning. It can happen that a client issues very specific instructions, but upon opening a dialogue there is clearly a misunderstanding about the basis of value or some other matter crucial to the whole exercise. Therefore, as with any other type of valuation, instructions need to be clarified and conditions of engagement need to be issued early on to avoid later confusion.

In this case study, the instructions were received from a bank specialising in asset-based lending. The bank was considering a loan to a company that manufactured plumbing products; the company was offering its plant and machinery assets as security for the loan.

The bank needed an assessment of the value of the assets to establish what funds it could safely advance to the business. Typical of the nature of this exercise, and indeed of many plant and machinery valuations, was that the detail initially provided was sketchy, although what was clear was that the deadline was crucial and the whole exercise was to be treated with a high degree of confidentiality.

Through discussion with the bank it became clear that what was required was a 'worst case scenario' valuation. This is typical of a valuation for this purpose. The bank needed to know how much the machinery could be sold for in the event that the borrower failed to repay the loan. Although businesses can be and are sold as complete operating entities, the bank did not require a business valuation, nor indeed a valuation of all of the company's assets. The machinery was to be the security for the loan in isolation, therefore it was not appropriate in this case to value the machinery on the basis of its installed value or value to the business. In the event that the bank called in its loan, only the machinery held as security would be involved. In these circumstances, the machinery would have to be removed from the defaulting borrower's premises at some stage. Thus the valuation scenario was one that reflected that the worst had happened, and the bank was seeking to realise its security through the sale and removal of the machinery through some sort of disposal process.

The valuation figures resulting from an exercise such as this are expressed in gross terms, which is to say exclusive of costs of sale. However, the bank specified that they wanted to calculate a net figure. Therefore, the anticipated costs associated with advertising and organising a sale, together with the further costs that would be incurred, were to be estimated so that the lender would be able to deduct them from the overall realisation. Thus the bank would be provided with a 'bottom line' figure to let them know exactly how much they could expect to receive from the security they were holding. They also instructed that the valuation should be based on the assumption that there would be a period of six months available to achieve realisation of the security at the indicated level.

The scope of the exercise was refined through further discussion. The valuation was to be limited to the major items of machinery that were considered suitable loan security. Although there was a subjective element to this stipulation, it was agreed that minor items such as office equipment and furniture, works furnishings and small tools and equipment were to be excluded. This fitted in with the valuation scenario: if it became necessary to realise the security for any reason then items of this nature would be difficult to identify and moreover would be of no great significance in the overall context of the valuation. Although it did not apply in this case, it would also have affected the treatment of plant that was 'built-in' to such an extent that removal would have rendered the equipment worthless or of minimal value. An example of this might be a cold room or large oven, where the installation costs represent a large proportion of the value of the asset: that value cannot be realised if the asset has to be removed so it does not represent good loan security. Further exclusions that are typically applied extend to dies, moulds and special tooling, as well as machine control programmes. In this case these exclusions were particularly relevant as it became apparent that the company had a very large quantity of tooling that was used on a number of the machines. The tooling in question was designed to produce a product to a specific design and could be used for no other purpose. Tooling of this type has no market value except scrap value unless the right to manufacture the product is included in its sale.

Description of the assets

We attach a schedule of the plant and machinery forming the subject of this valuation at Appendix 1. The assets comprise two production facilities engaged in the manufacture of brass and copper plumbing fittings, and briefly include ...

While the notes compiled by the valuer are usually incorporated in a detailed schedule or inventory attached to the report as an appendix, this report incorporated a general description of the assets and the processes involved in its main body. This provided an opportunity to comment on the nature of the machinery, its function in the manufacturing process of the company, and most importantly, its appeal to potential purchasers on the open market. As the report was addressed to bankers who could not be expected to have any background knowledge of the machinery or industry in question, this information was felt to be relevant in assisting the lender to assess the viability of the loan. It would not have been necessary if the report had been addressed to the operator of the machinery since obviously they would then be familiar with it. The following is a synopsis of what was written in the case study report.

The first facility produced a range of screw thread, compression-type pipe connectors in brass, such as are sold in the major DIY stores in the UK. The raw material was brass rod. The first stage of production involved cutting the rods into short cylinders using automatic cut-off saws. The cylinders were then heated and pressed into the approximate shape of the fittings bodies using a variety of standard power presses equipped with the appropriate tooling. In a secondary process the rough forgings were clipped of excess material by passing through a number of smaller power presses. Various machining operations followed using a large number of standard multi-spindle lathes, CNC lathes, machining centres and specialised rotary transfer machines. These operations bored the inside of the rough forgings to a close tolerance, faced the ends and cut threads on the fittings. Further operations involved cleaning and degreasing in barrel finishing lines, and assembly and packing using standard machines. The whole operation was on a reasonably large scale: there were 10 saws, some 80 various power presses, about 20 multi-spindle lathes, a similar number of CNC machines and rotary transfer machines, and three cleaning and degreasing lines.

The second facility produced ranges of plain and pre-soldered pipe connectors in copper. The raw material was copper tube and the primary process involved cutting this into appropriate lengths using automatic tube cutters and cut-off saws, and then annealing them (a heat treatment process to permit bending without over-stressing the metal) ready for secondary production. In the secondary processes the tubes were formed into the appropriate elbow, tee and straight coupler shapes using specially adapted power presses, and the bores and ends of the fittings were finished to the correct tolerances with specialised, purpose-built machinery. Pre-soldered fittings were produced in further processes employing specialised machines to bead the fittings and fill them with solder. The finishing operations of cleaning and degreasing, assembly and packing were carried out as at the brass production facility. Most of the machines were highly specialised, high production machines, and were relatively few in numbers of each.

Inspection

The plant and machinery included in this report was inspected between XXth and XXth January 200* by D M Kilbride MRICS. We confirm that he has relevant experience and knowledge to value machinery of this type.

This short report paragraph refers to the groundwork in any plant and machinery valuation. In itself, the task is a relatively uncomplicated one: it is simply a matter of gathering enough information about the subject-matter to be able to form an opinion of its value on the basis instructed. This information is the source material used by the valuer in carrying out the investigations necessary to come up with a figure. Typically, the information gathering takes place in three stages: an initial walk round with personnel at the factory to gauge the exercise and gain an overview of the assets and the processes involved; a detailed inspection where each machine is looked at individually; and a final enquiry stage to clear up any uncertainties about the equipment and to augment and supplement the information collected with input from the engineering and financial personnel of the business.

In this case, the initial walk-round revealed the scope of the exercise and it provided the opportunity to plan the work. Other general points that merited consideration at this stage included the method of removal of the machinery. The basis of value in this case assumed the machinery would be sold by some method, and therefore it was necessary to consider factors that might have an influence on this. Typical considerations to be addressed at this stage include the geographical location and the physical nature of the premises. Is the factory remote from areas where other potential users of the machinery are located; is it a multi-storey building; is there a roller shutter door opening onto a loading dock to facilitate easy removal; would demolition have to take place to allow a machine to be removed? Any or all of these factors have great potential to impact on realisable value, and thus require consideration from the valuer. In this case, the locations were in industrialised areas of the UK in easy reach of motorways, and the premises were conventional single storey industrial buildings with good access.

The detailed inspection will usually take the form of an individual assessment of each machine, and this starts by noting the information from the manufacturer's plate. Specifically, the manufacturer's name, their designation of model or type and the name or function of the machine will usually be required as a minimum, while the serial number can provide definitive identification when contacting a manufacturer at a later stage. A serial number is a manufacturer's unique identification number. In some instances, the serial number also carries information relating to the age or particular type of the machine. In this case, there was an issue of identification underpinning the work: with the knowledge that the bank was considering taking a charge over the assets, there was a necessity to eliminate as far as possible any potential future confusion over their identity. In practice, this meant recording any unique identification numbers: serial numbers were the ideal, but any plant, asset or maintenance numbers painted, riveted or stuck on the machines were recorded in addition.

Where no manufacturer's plate is visible, questions addressed to engineering personnel, or a trawl through their records, can prove fruitful. Even an experienced valuer will often be at a loss to know what a machine does or what to call it: operators can be useful sources for information like this, or else a manufacturer's handbook kept on or near the machine may be of assistance.

Further detail is also usually required, particularly with regard to age and sizes or capacities. Experience is the best guide to what information is required to form an opinion of value, and that is essentially the aim of a typical site inspection. Furthermore, it is well to stand back from the machine and form a general impression of its condition. The point of this is really to judge its appeal to a potential purchaser. Although a valuer is not required to make a technical survey, he should still attempt to judge how a purchaser would view the machine. Factors relevant to this include its age, its apparent standard of maintenance, its cleanliness, as well as the degree of ease or difficulty that may be entailed in removing it.

The final enquiry stage is hopefully a matter of gaining answers to questions that have arisen during the detailed inspection, and supplementing the information gathered so far with further detail or background, such as ages of equipment or the interpretation of asset registers, etc. This is most often achieved by interviewing local personnel, although plant and machinery valuers will have experience of situations where assistance of this kind is either withheld or given very grudgingly. In any case, information gained in this way needs to be treated with the appropriate amount of caution. During the course of the valuation forming the basis of this case study, time was extremely limited but engineering personnel were helpful. Of course, it was in the company's interest to be co-operative since the purpose of the exercise was to allow them to borrow money. A copy of the company's fixed assets register was made available, although detailed examination of this document at a later stage suggested the information it contained needed to be treated with a certain amount of circumspection.

Investigations

Where possible our valuations are based upon available evidence, derived from sales of similar machinery conducted in a free and open sales environment, adjusted where appropriate to reflect age, condition and other relevant factors. In the case of more specialist machinery that seldom or never changes hands, comparable evidence is difficult or impossible to obtain. In these cases our enquiries of dealers and manufacturers have been supplemented with depreciation techniques, taking into account age, condition, physical and technical obsolescence and other relevant factors.

The paragraph above describes very succinctly the two main approaches adopted in plant and machinery valuation: direct comparison and depreciated replacement cost. The choice of method depends on the evidence available, and indeed they are often used in tandem, weighted according to their relevance. A third approach occasionally adopted is income analysis, but this is a method more suited to the valuation of complete purpose-built plants and was not employed in this case.

Very briefly, direct comparison is a technique dependent on using knowledge of the price paid for one item to assess the value of a similar item. The method is reliant on having evidence of comparable transactions and is most applicable where the items are identical or very similar and transactions are numerous. The skill of the valuer lies in understanding how the difference between items affects value: as the degree of similarity between the items diminishes, increasing importance is placed on the subjective judgment of the valuer.

Depreciated replacement cost is a technique used when comparable market evidence is either absent or so rare as to make unreliable whatever evidence of transactions is available. Basically, the method is dependent on assessing the new cost of a machine, or its nearest current equivalent, and then deducting allowances for various factors so as to quantify the depreciation in value of the machine from its cost when new. The factors include the age of the machine, the type and amount of its utilisation, its operating environment and standard of maintenance to which it has been subjected, and the degree of obsolescence that has already affected it or is likely to affect it. Obsolescence can be product related, whereby what the machine actually makes or does is no longer economically viable in comparison with alternatives, or functional, whereby the way the machine operates is no longer efficient or productive compared with alternatives. The relevance of the factors in causing depreciation varies according to the type of machine, but it is important to note that however they are applied to a new cost, the resulting depreciated replacement cost is an economic value only: it is basically a measure of the remaining service potential of the item. Therefore, the technique is at best a rough guide that can assist in deriving a market value. When using depreciated replacement cost for this purpose, it is essential to reflect also on the possible level of demand for a particular item, or its appeal to a purchaser. In other words, it is important to try to envisage what someone might be willing to pay.

The automatic saws, tube cutters and CNC lathes and machining centres were general purpose machines that could be utilised in almost any type of engineering facility, while the packing machines and general items such as forklift trucks could be used in many industries. Similar machines commonly change hands in the second-hand market. Thus the first stage in the valuation process was examination of a database of auction realisations to obtain evidence of transactions in machines of similar age and capacity. This revealed a fairly consistent level of realisation for the various types of machine. 'Fairly consistent' is probably the best that can be hoped for: fluctuations always occur from sale to sale for a variety of reasons, not least being the fact that for all but the most common items, the number of purchasers do not generate a high enough level of demand to support a consistent level of value in the way that the market for cars or houses can. In practice this means that at any given time, only the slightest eagerness or indifference from purchasers can cause realisations to vary up or down from an accepted range that might properly be called a typical market value. Thus an auction database can only provide an approximate benchmark value for a certain type or model of machine. The benchmark in each case was adjusted according to the observations made during the site inspection: a machine that had stuck out as appearing to be above or below average condition was judged likely to attract a slightly higher or lower realisation, as the case may be. In some instances, the machines were almost brand new and while comparables were available of sales of older machines, there was no direct evidence for such recent machines. In these cases, manufacturers were contacted to obtain new cost prices for similar items. Since it seems fair to assume that no purchaser will pay more for a used machine than a new one (or else they could be classified as a special purchaser and would be excluded from consideration under the terms of the basis of value anyway) the comparable evidence for sales of older machines and the costs for new ones suggested the lower and upper limits of the range into which the subject machines would fall. Exact placement within the range was a matter of judgment, but factors that needed to be considered included the offering of warranties with new machines. Since the machines being valued would not have such warranties, any purchaser would be assuming an element of risk that would be reflected in the price they might be willing to offer. Observation of the markets in used machinery reveals that machinery in general loses most value in the early stages of its life. The level of demand and the current state of the market were also relevant factors, and were assessed in discussion with various dealers.

The most numerous machines involved in this exercise were power presses. These are basically machines that press material into a particular shape using a mould or die. All the machines were of British manufacture, all were made between the 1950s and the 1970s, and all ranged in size between 30–200 tonnes. Age is frequently the most important factor affecting value, but with machines of this age, values tend to 'bottom out' so that there is very little difference at either end of the age range. Machines of this age and capacity are very common on the second-hand market and can be readily bought at auction or from dealers. Thus the database of auction realisations was again consulted to obtain evidence of transactions in similar machines, and the comparable information was adjusted to reflect the machines actually being valued. Some machines were equipped with various extras such as automatic loading systems that were judged to make them significantly more attractive. However, the overriding consideration that most affected the valuation of these presses was their sheer number. It was suspected that there was a likelihood of flooding the market if all of these machines were to be released at once, and opinions of dealers specialising in power presses were sought to confirm or deny this suspicion. The detailed market knowledge of these specialists confirmed that demand was poor, values appeared to be falling and there was already a surplus of machines available. Therefore the release of some 80 machines onto the market at once would mean that flooding the market was more of a certainty than a likelihood, so a further adjustment to the benchmark market value had to be made. This adjustment could only be a subjective judgment, since there was little market evidence available to quantify it. In practice it was assumed that in the disposal scenario envisaged (a six-month timescale had been specified by the lender) not all of the presses could be sold. Therefore some of the poorer presses were valued at scrap value only, while the values of the others were revised downwards to reflect the sale scenario and the current state of the market.

Valuation of the multi-spindle lathes followed a similar process, but with a twist. These machines are designed to produce simple cylindrical components to a close tolerance in high volumes, and in the subject facility they produced small parts for various valves. The 20 or so machines were all six-spindle machines in the 1"–2" range (multi spindle lathes usually have 5, 6 or 8 spindles, the diameter of which ranges from less than 1" up to 6"). While the ages of the machines could be established from the manufacturer's plate as being between the early 1970s and mid-1980s, age was largely irrelevant, as the company had pursued an in-house rebuilding programme. Although lathes of this type are no longer available new, their reliability and durability means they are still in demand by users who can operate them efficiently and economically, despite the availability of more technologically advanced modern machines. The design of the machine lends itself to re-manufacture, with a frame that is extremely heavily built and not susceptible to significant wear. By replacing the moving parts around it and all the electrical controls, a machine can be obtained that is equal or superior to the lathe when it was new. It is therefore a feature of the market for these machines that there are a number of companies who buy old machines to refurbish before selling them on. Reference to a price database suggested age did not appear to be the overriding factor affecting the value of these items, but that the market was fragmented according to the nature of the purchaser. Heavily used, worn machines tended to command a level of value that reflected the appeal of the machine to those who rebuilt them whatever their age. Machines that retained some significant portion of working life because of light utilisation or a previous rebuild appealed to users and commanded higher values. Consultation with the relevant specialist companies revealed the typical prices that might be achieved for a rebuilt lathe, and it became apparent that only certain specific sizes could be sold easily due to market demand. However, at the time the market in general was very poor: rebuilt multi-spindle lathes had traditionally commanded good prices in the USA, but that market had fallen completely, with no prospect of immediate recovery. This meant the machinery rebuilders already had stock levels that they considered too high: they would have little interest in acquiring more machines, so there would be very little demand for the older, more worn subject machines. While some of the more recently rebuilt subject machines would be of interest to users in the UK and Europe, there would again be an expectation that realisations would suffer because demand would be swamped if all of the machines were made available for sale at once. Additionally, the machines rebuilt in-house would not be as well regarded as those coming from a company with a known reputation and expertise for rebuilding: there would naturally be caution about the in-house skills, and in any case the refurbishments had taken place over a number of years, so that the benefits brought about by rebuilding were not consistent across the range of machines. Thus valuation relied on careful weighting of a number of factors, all of which were felt to be influences on potential purchasers.

The rotary transfer machines load a part automatically, then transfer it to a number of stations, each one of which is set up to perform a specific operation such as drilling, tapping or thread cutting. As such they are fairly product specific, and potentially of limited appeal to other users. The machines were of Italian manufacture, and ranged in age from 1995–2000. There was no record of any recent transactions on the auction database. In addressing these machines, the initial impression had been that they were specialised: investigation sought to corroborate or refute this. Contact with the UK agent for the manufacturer revealed that the machines are designed for use in the brass fittings trade and although when they are supplied new they are individually configured to a particular customer's requirements, they use standard components that can be easily and cheaply re-configured to suit another user. The agent also advised that there were an average of 10–15 sold annually in the UK, with a total of perhaps 300 in the country, and that there was a large pool of users worldwide, with many machines being sold to India and China. Thus it could be inferred that there was a small market for these particular machines in the UK, and a greater potential for selling them overseas. The agent supplied the new costs for the machines, and also offered the name of a UK dealer. The dealer had sold similar but older machines recently. He confirmed that overseas purchasers in particular were interested in acquiring second-hand machines, although he was unsure of what price the newer ones might sell for. Even if he had offered an opinion on this, it would have been treated with a degree of circumspection, for the same reason that evidence provided by multi-spindle lathe dealers was: a specialised dealer offers added value in expertise and guarantees, and can therefore obtain higher prices than those that could be realised in the sort of sale envisaged under the valuation scenario. Armed with the knowledge of the approximate price that older machines might be expected to sell for, and coupled with the known new cost of new machines and the reasonable assumption of some demand, a judgment could be made on the value of the subject machines. This was again adjusted according to the ages and conditions of the machines observed during the inspection.

There were a number of specialised machines where direct comparison would simply be impossible due to lack of evidence: these included the specially adapted forming presses and the purpose-built facing, sizing, and solder filling machines at the copper fittings facility, and the cleaning and degreasing finishing lines at both facilities. These machines were initially valued on a depreciated replacement cost basis. The new cost was established by contacting manufacturers, or in some cases where this approach proved unsuccessful, by indexing historic costs to allow for inflation. Inflating historic costs is really an approach of the last resort, since there are many factors besides inflation that can influence the price of new machinery, such as improving technology or changing markets. As so much machinery is purchased from overseas in today's global economy, there is also the issue of fluctuating exchange rates. However, when machinery suppliers are no longer in business or refuse to divulge information, there is sometimes no other option. The typical economic working life of each type of machine was then estimated. This may seem like crystal ball gazing, but in practice is relatively straightforward when some experience has been acquired. The key is that a typical, rather than specific, lifetime is being predicted. While an individual machine may be in use many years after its assessed economic life is over, or indeed scrapped many years before, in general it can be observed that various types of machines have average working lives, for instance perhaps four years for a personal computer, or 10 to 15 years for a machine tool, or 20 to 25 years for a power press. By comparing the actual age of the machine with its anticipated life, a factor to allow for depreciation from new can be derived: most plant and machinery valuers refer to a valuation table to derive the factor. This factor is then applied to the new cost to produce a measure of economic value. However, the value thus calculated is an approximate indicator only: it is not a market value and does not allow for supply and demand. Therefore to actually come up with a valuation for these special purpose machines, further investigation and consideration was necessary.

In the case of the forming presses, the adaptations had been designed in-house by the company and implemented as costly variations to standard machines by the manufacturer that made them very efficient at making the copper pipe fittings, but unable to produce anything else. Due to the specialisation of the machines it was felt that they would not appeal to dealers, and the market would be limited to just two groups of purchasers; first, manufacturers of similar products, and, second, other users willing to undertake the cost and risk of adaptation. The first group was known to be few in numbers and would already have their own tried, tested and trusted machinery. In the case of the second group, while the machines could be re-tooled to make some other product, a potential purchaser would take the cost of readapting the machine into account, and would adjust any price they might be willing to pay to reflect this. In either case, while there can be no doubt that users of machinery can be tempted to buy if the price is low enough, the particular circumstances here meant that due to the age, specialisation and relocation costs of the items the price would have to be very low to offer any incentive. Therefore there were good grounds for valuing the machines at something significantly below depreciated replacement cost.

Enquiries addressed to the company concerning the facing, sizing, and solder filling machines revealed that these were bespoke items, built specifically for the company and to their design. The depreciated replacement costing of these items followed the normal lines, although assessment of new cost and economic working life was carried out in conjunction with engineering personnel and required careful consideration and thorough research of historic costs. The market for these items would be limited just to producers of similar fittings, since they could not be economically adapted for anything else. In the case of the solder filling machines, some industry research revealed that the fittings produced by these assets are only used in the UK market, further limiting the number of potential purchasers. As the machines were all of considerable age, it was considered that the chances of selling any were small. Consequently, most were valued at little above scrap value.

The degreasing and cleaning lines had been supplied as specifically dedicated plants using various standard components. One of the lines dated from the early 1980s, the other three were early-to-mid 1990s. The manufacturer's agent supplied new cost information, and advised that alternative uses for the lines were limited, although the newer lines could be converted to a more general application at a price. He also estimated that the typical working life for a line in its current application was around 20 years. The depreciated replacement cost for the older line was therefore already low, and it was felt that given the costs of relocation, it would be unlikely to attract a purchaser as a complete line, although certain components could probably be sold off individually. The newer lines were more attractive although again relocation costs would be high: in consideration of these costs and the anticipated conversion costs, a valuation significantly below the depreciated replacement cost was adopted.

Valuation

In our opinion, the value of the assets as detailed on the attached schedule on the basis of Open Market Value as at XXth January 200X is...

It often seems that this is the only section of the report read by the client.

Exit strategy

You have instructed us to advise you on the most appropriate method of disposal within a six months timescale. You have also asked us for an estimation of the costs that would be incurred to realise the security at the reported values. Our comments are as follows...

The lender had specified that the valuation be based on a realisation period of six months. If the period had not been considered long enough to maximise the realisation the report would have had to be qualified accordingly. In assessing an appropriate period, estimation must be made of the time required to allow suitable preparation of the site for a sale, proper marketing so as to alert a wide enough spectrum of potential purchasers, and a reasonable time in which to effect clearance. The factors that influence the estimate are largely related to who the target audience is, and how long clearance might take. In this case it was felt that many of the assets were adequately specialised and valuable to justify advertising on a European, if not worldwide, basis. Placing adverts in the appropriate trade publications sometimes requires a long lead time, particularly if they are published monthly or bi-monthly, while identifying a target group, purchasing or compiling a database of their names and addresses, and undertaking direct marketing can be similarly time-consuming. However, with a six-month timescale it was felt that an adequate period was available to alert all potential purchasers to the availability of the items. Clearance was not really a major issue in this case, as it was believed that almost all of the items could be removed within a matter of weeks. However, if the items had been of a particularly complex installation, or had required demolition (and subsequent reinstatement) of the buildings, it might have required more careful consideration.

The most appropriate method of sale is really a question of the nature and number of the assets. To achieve a good realisation at an auction it is necessary to attract a large enough number of willing buyers to produce competitive bidding. It is accepted that to do this it is necessary to have a reasonable number and mix of lots of general appeal. An auction sale of only a half dozen old personal computers would not be cost effective. Similarly, a food processing machine will probably not realise well if it is the only such machine in a sale of numerous woodworking machines: it would simply not be worthwhile for interested parties from the food industry to attend in the knowledge that only the one machine will be available to them. Where the items are more specialised or fewer in number, there is a risk that not enough potential purchasers will attend on one specified day so that a sale by tender is probably a more effective option. This method extends the bidding period, so that it is not necessary to have all potential purchasers in one room at any given time. It also requires those who tender to offer the maximum that they are willing to pay rather than simply beating the last bid. Where assets are very specialised or valuable, sale by private treaty may be more appropriate since this approach facilitates negotiation and careful consideration of any offer.

The advice to the client in this case was that all three methods of sale might be appropriate for different machinery. It was recommended that the more specialised items such as the copper forming presses, the sizing, facing and solder filling machines, and the cleaning and degreasing lines be initially offered for sale by private treaty, with the hope of attracting a purchaser who would use them in the same industry. If such a purchaser could be found, they should be willing to pay more than someone who would have to convert them for an alternative use. If sale by private treaty was unsuccessful, the machines could then be offered in a sale by tender, which might also incorporate other fairly specialised machines such as the rotary transfer machines. For cost effectiveness the tender could be advertised and organised alongside an auction for the more general purpose engineering machinery. Sale by auction was considered the most efficient method of disposal for the saws, power presses, lathes and other more general-purpose machines since their number and quality would be capable of generating a good level of interest.

In estimating the costs of sale, regard was had to the typical expenses that would be incurred. Porters would be required to tidy up the site and the machines, organise the various lots into a sensible order to enable easy inspection, and to provide general assistance during viewing, selling and clearing periods. In addition, it is often useful to employ maintenance or production staff from the subject company to assist, since they can supply technical expertise and may be able to demonstrate machines under power to interested parties. Professional staff would be required to oversee these activities and compile a catalogue, as well as actually conduct the auction and manage the whole sale. Administrative staff would be required to assist with invoicing and banking. Security guards would be required on viewing and sale days in view of the anticipated size and nature of the sale. In each case the number of man-days was estimated, and multiplied at the appropriate rate. Additional expenses that were included in the overall figure advised related to professional photography, purchase of a database for target marketing, printing of a brief 'flyer' highlighting the main items and of the full catalogue, postage costs for a mail shot of the flyer and the catalogue, and advertising costs in the relevant trade journals. In this particular case, the premises housing the plant and machinery were freehold. If they had been held on a lease, it would have been necessary to examine the lease to ascertain that a sale of the nature intended could actually be held, and to calculate the amount of rent payable. As it was, occupation costs were assessed in conjunction with financial personnel of the company, to include rates and utilities that would be payable, albeit that the latter would be at a reduced rate. All of these costs would be borne by the lender in the event that they needed to realise the security they held under the loan, and were therefore relevant to their assessment of the sum they were willing to advance to the company.

Exclusions

We have excluded the following from our valuation:

(i) *The land and buildings along with the occupier's fixtures and fittings including overhead cranes.*

(ii) *Raw materials and consumables, work in progress and finished goods.*

(iii) *Goodwill, patents, trademarks brand names and other intellectual property.*

(iv) *Dies, moulds, patterns, jigs, fixtures, tools, machines control programmes, special tooling and other items which are product dedicated rather than of a general application.*

(v) *Computer software, licences, tapes, discs and data, drawings, designs, technical data, and commercial and administration records.*

(vi) *Third Party Assets. These in the main comprise of contract rental vehicles, forklift trucks and some office equipment.*

These are mostly self-explanatory general exclusions that are incorporated into all reports, although they need to be agreed with the client beforehand and reference to them included in the Conditions of Engagement letter. The reason for some of them is simply that they are outside the plant and machinery valuer's sphere of expertise: buildings, stock and goodwill are the province of other specialists. The relevance of the exclusion of press moulds and tooling in this case has already been referred to. There is an argument that items such as these are in fact part of the goodwill of the business because they are inextricably linked to the product and therefore the profitability of the business. More importantly, this case concerned only the plant and machinery of the company: it was perceived that in this circumstance the tooling would have no realisable value except as scrap, and therefore would offer no significant security to the lender. The point applying to the exclusion of third party items is also relevant: obviously a lender cannot expect to sell the property of others to recover a debt owed to it by a borrower.

Title issues

We have been advised that all the assets referred to in our report are owned by the company, have clear title and are free of charge and unencumbered. No provision has been made in respect of any retention of title claims upon goods supplied and for which values are contained in this report.

There is an increasing trend for manufacturers to release the capital tied up in their equipment by using it as security for finance. This was obviously the purpose of this valuation, so at an early stage it was appropriate to ask management at the subject company whether there were already any financial agreements in place with other lenders. In some valuation scenarios it is appropriate to value assets held under lease purchase agreements. For instance, in a valuation of an insolvent company the value of such assets may be higher than the debt outstanding under the agreement. Therefore the assets might be liquidated to realise their equity. In this case if any assets had already been pledged to another lender they could not have been taken as security for the loan envisaged, and would have been excluded from the valuation. The sentence on retention of title is a reference that is more usually applied to stock valuations. It is common practice for suppliers to insert a clause into their sales agreements to the effect that title to the goods does not pass until payment is received in full. This is a protection against the customer becoming insolvent: in that event the supplier may be able to repossess the goods if they are not paid for. Its relevance in a report such as this is limited, but if a machine had been recently delivered it may have been possible that full title had not passed at the time of valuation. The sentence makes it clear that this possibility has not been investigated.

Caveats

As with most valuation reports, a number of caveats are inserted to frame the context of the exercise, to clarify any potential misunderstandings, and to prevent misinterpretation of the opinion expressed. While some of those contained in a plant and machinery report are common to other valuation reports, the following are reasonably specific and merit further comment.

The values set out in this report must be considered as a guide to the value of the assets as a whole and in the event of individual items being sold, then the value of the remaining assets may be substantially affected. We do not recommend that individual values be utilised to calculate loan collateral.

This emphasises the point that the assets are valued as a package. If the piecemeal selling off of individual items changes the mix of the package, it may be detrimental to the remaining items. This is particularly relevant if the major items are sold. In an auction sale scenario, for instance, the major assets tend to draw in potential purchasers who will hopefully also add to the body of bidders for smaller items. If the major items are not available, the appeal of the remaining assets is lessened. In this case, the caveat was a pointer to the bank that 'cherry picking' particular items out of the valuation might not offer them the level of security that was indicated by the individual assessment of that item because of this tendency.

Our valuation disregards the possibility of a sale to a special purchaser. Accordingly, where appropriate, careful consideration should be given to an asking price, as it is possible that a purchaser with special needs may be prepared to pay a figure in excess of our valuation.

The valuation is an assessment of the market value of the items. This assumes that buyers and sellers operate without compulsion. It can happen that a particular individual has a particular need for a piece of equipment that compels him to offer a price above the market value. Since this cannot be predicted or quantified, it is excluded from the assessment. This caveat alerts the client to this possibility in the event that a sale has to take place.

We have not carried out any tests upon the plant included in this valuation. We have assumed that all items have been maintained and are fully operational in accordance with the manufacturer's recommendations and that they comply with all statutory and Health and Safety requirements.

A plant and machinery valuation is based on a physical inspection, but that inspection is not a technical survey. Although it may be appropriate to make specific enquiries about old or redundant equipment, or equipment that is known to be affected by impending changes to regulations, the assumption that equipment is operated in accordance with current legislation is reasonable, since it should not be operated otherwise. If a piece of equipment were not so operated, for instance if a machine was not fitted with suitable guarding, it is likely that any realisation would be depressed since the purchaser would be obliged to pay to bring the machine into line with current standards.

Conclusion

The case study presented above is based on an exercise actually undertaken, although details have been changed or omitted to protect the innocent, and various embellishments have been made to illustrate the main approaches adopted. For reasons of brevity and clarity, a great deal of detail has been omitted, but it can nevertheless be seen that there is a large amount of research and analysis behind a typical report.

While the approach adopted in any case may have certain aspects that are typical, each case will raise different considerations, and the necessary research, analysis and enquiries will vary accordingly. This is true even of valuations on the same basis because almost all industrial facilities are unique. The range of purposes that valuations might be requested for has not been addressed here, because the case study is of one particular exercise. However, that range of purposes is large, and this adds even further to the diversity of the work. It is perhaps this aspect of plant and machinery valuation that makes it most interesting and challenging.

Further reading

For those wishing to find out more about plant and machinery valuation in the UK, there is a definite lack of authoritative source material.

Valuation of Plant and Machinery by Chris Derry, (College of Estate Management, 1985) is clear and thorough, but it is now out of print and dated in some aspects.

The RICS *Appraisal and Valuation Manual* ('Red Book') sets the parameters we work within, but is silent on most practical matters. Apart from this, the body of writing in the UK devoted to plant and machinery valuation is limited to occasional specific journal articles.

An international perspective can be gained thanks to the more prolific output from valuers in the USA, although their approach and terminology can vary quite significantly from ours.

Valuing Machinery and Equipment, published in 2000 by the American Society of Appraisers and written by their Machinery and Technical Specialities Committee, is an extensive reference work and is no doubt essential reading for appraisers in the USA.

Appraisal Principles and Procedures by Henry Babcock, (5th edition published in 1994) is a more general, more theoretical work.

Dimensions of Value by Leslie Miles *et al*, (6th ed published 1997 by MB Valuation Services Inc.) offers many practical examples of plant and machinery valuation in the USA.

Valuation of Plant & Machinery (Theory & Practice) by Kirit Budhbhatti (published by Kirit Budhbhatti in 1999) attempts to gain a fully international perspective, but is closely focused on India.

Chapter 10

Giving Expert Evidence

The Royal Institution of Chartered Surveyors has published a document entitled *Surveyors Acting as Expert Witnesses – Practice Statement and Guidance Notes*. It obliges all chartered surveyors who undertake this role to adhere to it. Those so doing who are not chartered surveyors should however read it and are advised to follow it. This chapter sets out suggestions in respect of other aspects of litigation with which valuers may be concerned; there is inevitably some overlap. When a witness is to give evidence in a Court of Law (as distinct from an administrative tribunal or before an arbitrator) he must comply with the Code of Guidance on Expert Evidence published by the Lord Chancellor's Department and his report or proof of evidence must contain a statement that it has been complied with; other similar documents are the Academy of Experts' Code of Guidance and the Law Society's Code of Practice for expert witnesses engaged by solicitors.

Expert evidence is opinion evidence which can be contrasted with factual evidence. The word 'expert' is sometimes used where 'specialist' would be more appropriate. That one building is taller than another is a matter of fact: that a building is worth £100,000 freehold is a matter of opinion, expert evidence. An experienced valuer who has practised in a town for some years is probably capable of valuing a shop in that town for which purpose he can give expert evidence, ie state his opinion of its value. Another valuer practising countrywide who does little else but value shops is better described as a specialist although he is also an expert for the purpose of making a valuation of the same shop. An expert need not necessarily have any formal qualifications although possession quickly establishes him as an expert. Evidence given by a valuer is almost always a mixture of expert and factual evidence.

The courts and tribunals where a valuer may give evidence include: the courts (High Court, County Courts, Magistrates Courts, and Crown Courts). The Lands Tribunal, Rent Assessment Committees, Valuation Tribunals, Leasehold Valuation Tribunals, Agricultural Land Tribunals, Planning Inquiries and Appeals and Arbitrations.

A valuer should aim to keep his client out of court and obtain a proper settlement by negotiation (if he is instructed to negotiate) but should not avoid going to court or tribunal just because it is an onerous task. During negotiations he should be constantly asking himself: 'Should this matter go to court' and the answer may vary with time as the matter progresses. Similarly at all stages, the valuer negotiating should ask himself: 'Have I failed to communicate my case to the other side?' It may be well worth while to take the initiative by instituting proceedings: a reference to Lands Tribunal by one side often has a remarkable effect on negotiations where the other side will not move. Once taken, the initiative should not be lost: applications for extension of time should be resisted pressure should be kept up.

Sometimes a valuer may receive instructions 'out of the blue' in respect of a property with which he has not been concerned before: his initial reply should be to the effect that he will inspect and report.

Sometimes the valuer comes into the matter indirectly; for example, if he is managing a property. He should not agree to appear as an expert witness if he considers there is no case or if the matter is outside his experience, although with new legislation, it may be that no one has any experience of that matter. Instructions offered close to a hearing date already fixed present particular difficulties. The obvious solution is to try to get a postponement of the hearing but failing this it must be made absolutely clear to all concerned (the client, the lawyers and the bench) that preparation time has been too short for full evidence to be given.

From time to time, the valuer is faced with the problem of a client who is determined to have his day in court, 'to have a go at the other side.' The valuer should resist becoming involved and should never try to help the client, who is determined to go to court by agreeing to give evidence: – it is the valuers' opinion which is required, not the client's view. There are, however, circumstances where the matter should be taken right up to the hearing: it has been put that sometimes the best time to settle is 'a minute before 10.30.'

As soon as it is decided that the matter is to go to a hearing and that it is one where counsel is to be instructed, the matter should be put before counsel as soon as possible. That is primarily a matter for the solicitor, as is the choice of counsel, but the valuer concerned might well suggest this to the solicitor. For the purpose of putting the matter to counsel it may be appropriate for the valuer to make a preliminary report to provide the material to form the basis for instructions to counsel. Such a preliminary report should be as complete as possible but at an early stage it may well contain headings without matter below, probably because there has not been enough time to make the necessary inquiries. It should also indicate the other side's case as far as it is known and the state of any negotiations.

As soon after the process has been initiated (for example a reference has been made to the Lands Tribunal) it will be necessary for the witness to prepare a proof of evidence or report to enable his advocate to know what evidence will be given by him. This will probably be similar to any preliminary report made for counsel but in full detail.

The first draft is usually written by the witness but the solicitor (and counsel) may comment on it. It has been known for the solicitor to alter the wording; the witness must resist this (and be very firm in so doing) if the altered text in any way does not express his or her opinion. Appendices may be useful and a separate document headed 'Notes for Counsel' may be useful in putting counsel completely in the picture. The witness should consider drafting an agreed statement of facts: should obtain instructions to do so (and what to agree) and to meet the other side's witness to this end. The object of this statement is to save time at the hearing; it is obviously a waste of time for the parties during the hearing to give contrary evidence in respect of a certain dimension for example when it could have been measured on site by both parties together. It is self-evident that the property the subject of the proceedings should be included in such a statement but so should 'comparables.' The witness should not agree that a transaction is agreed to be admitted in evidence without express instructions.

Thus he may agree that if a certain property let at a certain date on certain terms at a certain rent, the rent represents £x per m^2, but expressly that the transaction is not admitted in evidence. The importance of this distinction has diminished since the Civil Evidence Act 1995 which provides for the admission of hearsay evidence subject to provisions and conditions and to the service of notice. If the latter has to be given, it should be drafted and served by a solicitor or, if the Valuer deals with it, the notice should be read by a solicitor before service.

Chartered surveyors now have direct access to Counsel and the guidelines issued by the Institution should be strictly followed if this course is taken.

As soon as the valuer forms the view that the dispute is likely to come to a hearing, it may be good tactics to keep pressure on the other side by frequent letters, telephone calls, faxes and meetings. This is particularly important if the other side drags its feet.

The valuer should not hesitate to ask the solicitor for advice from counsel either by way of a written opinion or a conference with counsel. Whether this suggestion is accepted by the solicitor is a matter for him, not the valuer, unless the valuer himself is instructing counsel direct. If a conference is agreed, the valuer should attend (but again whether he does so or not is a matter for the solicitor) and if he does he should expect to take part in the discussion.

Certainly in a large case, there is advantage to be had if the following act as a team: counsel, the instructing solicitor, the valuer who is to give evidence and the client. The valuer should know the subject property intimately: it may be an advantage if the valuer himself measures it. It is essential that the valuer should view the comparables on which he relies and also those relied on by the other side, both lots internally if possible. Original notes made on all these inspections may prove to be very powerful evidence in court in the event of dispute.

Pre-trial reviews

This is a meeting held before a substantive hearing, the object being to arrange for the 'just, expeditious and economical disposal of the proceedings.' The usual representatives appearing are the parties' solicitors but sometimes counsel is instructed. It is called a 'pre-trial review' in the Lands Tribunal, a 'pre-hearing hearing' in a Valuation Tribunal, and a 'preliminary meeting' before an arbitration; there is a similar arrangement in the courts called a 'summons for direction.'

It is called by the bench concerned by notice being given determining when and where it is to be held. There is no set procedure: the matters usually dealt with are when and where the substantive hearing is to be held: an estimate of the probable duration of the hearing may be mentioned. The position as to agreement on facts is usually raised. This is an opportunity for applications to be made for example if extra expert witnesses may be called or an order for disclosure.

The result in the Lands Tribunal is an order which summarises the matters covered and determines who should pay the costs involved in holding the pre-trial review: this is usually 'costs in the cause', ie it follows the order for costs in the substantive hearing. A further pre-trial review may be ordered.

A pre-trial review should be distinguished from the substantive hearing: particularly where the parties (or one of them) are not communicating properly, it is a most useful device and often results in settlement or a narrowing of the issues in dispute.

The roles of those concerned (who does what)

In any usual case of any size or importance, a solicitor. Counsel and witnesses are involved: the role of each is not always known. It is the solicitor's task to see that notices are complied with and dates adhered to: the solicitor is the stage manager, whose job is to see that counsel has all the necessary papers. The solicitor should draft documents, referring to counsel if he considers it desirable. The solicitor should check any proofs of evidence or reports to be spoken to by witnesses. At the hearing he or she should keep a note of the proceedings. Counsel's task is to present the case, to call witnesses, to cross-examine and to make submissions including matters of law. While counsel may present the case in its best light (and put the opponent's case in its worst light) (s)he must not mislead the bench and that includes not disclosing a relevant matter. The expert witness is there to enable the bench to come to the right conclusion and not there to win the case.

Documents

The following is a check list which applies to *documents* to be put to the bench:

1. Strong paper, flat and legible.
2. Binding – not too tight nor too loose.
3. Reference number to be shown.
4. Sufficient copies plus a spare or two.
5. Cross-referenced to plan or to other documents.
6. Any schedules to have numbers or letters vertically and horizontally for each column.

The following is a check list which applies to *plans*:

1. To a professional standard.
2. NEVER rolled.
3. Reference number to be shown.
4. Scale.
5. North point.

6.   Key.
7.   One agreed plan is to be referred to one from each side.

The following check list applies to *photographs*:

1.   To a professional standard.
2.   To be agreed – if possible.
3.   Described and dated on front.
4.   Reference number.
5.   To be accompanied by a plan showing a point where taken and direction of camera.
6.   It assists if the same reference numbers are used for all three of the above.

Character of a hearing

The degree of formality may vary with the level. In the High Court the proceedings are formal, in Lands Tribunal there is less formality but the rules of evidence are kept to. In a local tribunal, great latitude may be given. Wherever the hearing is, it is not a review of negotiations and no reference should be made to them whether or not it was expressly stated at the time to be without prejudice. In the Lands Tribunal and the courts, the findings have to be made on the evidence adduced at that hearing. The position of a valuation tribunal, a rent assessment committee or a leasehold valuation tribunal appears to be that the decision is not so limited: see *Crofton Investment Trust Ltd* v *Greater London Rent Assessment Committee* [1967] 2 QB 955.

Sequence of a hearing

Subject to certain rules (for example in the Lands Tribunal in a matter concerning compensation for land acquired compulsorily when the claimant has the right to go first even if the reference was made by the acquiring authority) the party who took the initiative in the proceedings starts. The advocate opens his case by addressing the bench and puts it in the picture going through all or some of the documents. The advocate then refers to the law, statutes and case law, on which the advocate relies. The advocate then calls his witness who gives his evidence in chief. To save time his proof of evidence or report is sometimes taken as read. The advocate who called him may prompt him by for instance, saying, 'What about the approach road?' but he must not lead the witness by putting words in his mouth for example 'The approach road has a poor surface, no foundation and is 1 in 6 isn't it?'

At the conclusion of the witness's evidence, the advocate for the other side has the opportunity to cross-examine the witness. The rule then is that any relevant question (including leading ones) may be put: it is not limited to the evidence given in examination in chief. The object of cross-examination is to test the witness, not to trip him up although that occurs.

There follows what may perhaps be the most difficult part of the witnesses task – re-examination. The rule here is that only questions (and not leading questions) which relate to matters raised in cross-examination may be asked. Other witnesses may then be called and the same practice followed. That will conclude the case of the advocate who opened.

The other side then calls his witnesses who are similarly examined, cross-examined and re-examined.

The advocate who appeared for the responding party then has an opportunity to make his submissions and to deal with the law as he sees it, no doubt criticising his opponent's submissions.

Finally, the advocate who opened the case has the right of reply but he should not raise any new matters.

Without prejudice

Letters and documents so marked should, except possibly in very rare instances, not be referred to at any time during the hearing. The marking only has effect after a dispute has arisen, thus a letter from one person to another asking if the latter would sell a property would not be protected, no dispute having arisen. The principle behind this is to encourage settlement without a hearing.

'Calderbank' letters

This is a letter written by one party in a dispute to the other: it includes an offer to settle and sets out the terms. It adds words to the effect that the offer is made without prejudice but that the writer reserves the right to refer to the letter if and when the matter of costs was raised. It arises from the decision of Heilbron J in *Calderbank* v *Calderbank* [1975] 3 WLR 586. There are specific rules in the Lands Tribunal concerning unconditional offers, frequently referred to as 'sealed offers.'

Costs generally

In the courts and in the Lands Tribunal, each side may be involved in paying the costs of counsel, the solicitor and witnesses. Generally the bench has discretion as to who pays these costs: the losing party usually pays not only his costs but those of the other side as well. This is subject to specific directions in the Lands Tribunal in some disputes. Costs are awarded on scales for example county court or High Court. Costs may be taxed (by a taxing master or by the Registrar in the Lands Tribunal) that is they may be assessed in the event of a dispute between the parties, often as to the amount claimed. There is a well known adage in respect of costs: 'If you don't ask for them you may not get them.' The question of costs should be raised or reserved at the hearing and may be dealt with subsequently by written representations. There are no provisions as to costs in valuation tribunals or rent assessment committees.

Combining roles

To save costs in local tribunals it is not uncommon for a valuer to combine the roles of advocate and expert witness. This is both difficult and burdensome. It is essential to make it very clear to the bench when each role is being played. The two should never be mixed.

How to give evidence (In the box – some dos and don'ts)

1. Take all papers, calculator, scales etc. which might possibly be needed and have them *at hand*.
2. Take your time – neither hurry nor let an advocate hurry you.
3. Beware of the advocate for the other side putting you at your ease.
4. If you refer to a paper or file, very often the advocate for the other side may ask to see it.
5. Look at the advocate who is speaking to you but deliberately turn and answer to the bench.
6. When being cross-examined, it is sometimes difficult to know whether to add or not to your answer, a plain 'yes' or 'no.' Try to catch the eye of the advocate for your side and add nothing, he will pick up the point in re-examination if he thinks it should be pursued.

7. Speak clearly – treat the bench as being very slightly hard of hearing.
8. Keep to evidence: do not refer to the law.
9. Listen carefully to the question being put and answer it, some witnesses seem to catch just one word and say what they know about that subject, not answering the question at all.
10. If more than one question is put in one sentence, break it up and answer each separately. If you do not understand the question ask for it to be put again and/or in another way.
11. Keep answers short. Do not use slang expressions.
12. If you consider you can, show some flexibility in your answer. For example, if there is a figure of say, £10,000 in your calculation, be prepared to admit that it could be lower or higher.
13. Keep cool: do not let the cross-examining advocate rattle you.
14. Once you are called to give evidence remember you may not communicate directly or indirectly with the advocate who has called you.

© Bill Rees

Index

For Product Safety Concerns and Information, please contact
our EU representative GPSR@taylorandfrancis.com
Taylor & Francis Verlag GmbH, Kaufingerstraße 24, 80331 München, Germany

T - #0080 - 230425 - C0 - 216/138/12 - PB - 9780728204188 - Gloss Lamination